QCE Units 3 & 4
CHEMISTRY

BRETT STEEPLES

+ summary notes
+ revision questions
+ detailed, annotated solutions
+ study and exam preparation advice

STUDY
NOTES

A+ Chemistry QCE Units 3 & 4 Study Notes Workbook
1st Edition
Brett Steeples
ISBN 9780170459150

Publisher: Sam Bonwick
Editor: Marcia Bascombe
Project designer: Nikita Bansal
Permissions researcher: Liz McShane
Production controller: Jaimi Kuster
Typeset by: SPi Global

Any URLs contained in this publication were checked for currency during the production process. Note, however, that the publisher cannot vouch for the ongoing currency of URLs.

For product information and technology assistance,
in Australia call **1300 790 853**;
in New Zealand call **0800 449 725**

For permission to use material from this text or product, please email
aust.permissions@cengage.com

ISBN 978 0 17 045915 0

Cengage Learning Australia
Level 7, 80 Dorcas Street
South Melbourne, Victoria Australia 3205

Cengage Learning New Zealand
Unit 4B Rosedale Office Park
331 Rosedale Road, Albany, North Shore 0632, NZ

For learning solutions, visit **cengage.com.au**

Printed in China by 1010 Printing International Limited.
1 2 3 4 5 6 7 25 24 23 22 21

CONTENTS

UNIT 3

EQUILIBRIUM, ACIDS AND REDOX REACTIONS

UNIT 4

STRUCTURE, SYNTHESIS AND DESIGN

HOW TO USE THIS BOOK

A+ Chemistry QCE Units 3 & 4 Study Notes is designed to be used year-round to prepare you for your QCE Chemistry exam. *A+ Chemistry QCE Units 3 & 4 Study Notes* includes topic summaries of all the key knowledge in the A+ Chemistry QCE syllabus that you will be assessed on during your exam. Chapters 1 & 2 and 4 & 5 address each of the topics in the syllabus. Chapter 3 specifically addresses the Data test, which is a separate assessment task worth 10% of your final mark. This section gives you a brief overview of each chapter and the features included in this resource.

Topic summaries

The topic summaries at the beginning of each chapter give you a high-level summary of the essential subject matter for your exam.

Concept maps

The concept maps at the beginning of each topic provide a visual summary of the hierarchy and relationships between the subject matter in each topic.

Key knowledge summaries

Key knowledge summaries in each chapter address all key knowledge of the study design. Summaries are broken down into sequentially numbered chunks for ease of navigation. Step-by-step worked examples and hints unpack the content and prevent mistakes.

Glossary

All your key terms for each topic are bolded throughout the key knowledge summaries and included in a complete glossary towards the end of the chapter. Digital flashcards are accessible via QR code and provide a handy revision tool.

Revision summary

The revision summaries are a place for you to make notes against each of the syllabus dot points to ensure you have thoroughly reviewed and understood the content.

Exam practice

Each topic ends with multiple-choice and short response questions for you to test your recall of the key concepts and practice answering the types of questions you will face in your exam. Complete solutions to practice questions are available at the back of the book to provide immediate feedback and help self-correct errors. They have been written to reflect a high-scoring response. This section includes explanations of why the multiple-choice answers are correct, and explanations for short response items that demonstrate what a high-scoring response looks like, with mark breakdowns, and signpost potential mistakes.

PREPARING FOR THE EXAM

Exam preparation is a year-long process. It is important to keep on top of the theory and consolidate your knowledge regularly, rather than leaving revision to the last minute. Revise often, choosing one or two dot points to focus on. You should aim to have the theory learned and your notes complete so that by the time you reach study leave, the revision you do is structured, efficient and meaningful.

Study tips

To stay motivated to study, try to make the experience as comfortable as it can be. Have a dedicated study space that is well lit and quiet. Create and stick to a study timetable, take regular breaks, reward yourself with social outings or treats and use your strengths to your advantage. For example, if you are a visual learner, turn your Chemistry notes into cartoons, diagrams or flow charts. If you are better with words or lists, create flash cards or film yourself explaining tricky concepts and watch it back.

Revision techniques

Here are some useful revision methods to help information **STIC**.

Spaced repetition	This technique uses the Leitner method, which helps to move information from your short-term memory into your long-term memory by spacing out the time between when you are asked to revise or recall information from flash cards you have created. As the time between retrieving information is slowly extended, the brain processes and stores the information for longer periods.
Testing	Testing is necessary for learning and is a proven method for exam success. The 'hypercorrection effect' shows when you are confident of an answer that is actually incorrect, you are more likely to remember the correct answer, thereby improving your future performance. Further, if you test yourself before you learn all the content, your brain becomes primed to retain the correct answer when you get it.
Interleaving	A revision technique that sounds counterintuitive but is very effective for retaining information. Most students tend to revise a single topic in a session, and then move onto another topic next session. With interleaving, you may choose three topics (1, 2, 3) and spend 20–30 minutes on each topic. You may choose to study 1, 2, 3 or 2, 1, 3 or 3, 1, 2 'interleaving' the topics, and repeating the study pattern over a long period of time. This strategy is most helpful if the topics are from the same subject and are closely related.
Chunking	An important strategy is breaking down large topics into smaller, more manageable 'chunks' or categories. Essentially, you can think of this as a branching diagram or mind map where the key theory or idea has many branches coming off it which get smaller and smaller. By breaking down the topics into these chunks, you will be able to revise the topic systematically.

These strategies take cognitive effort, but that is what makes them much more effective than re-reading notes or trying to cram information into your short-term memory the night before the exam!

Time management

It is important to manage your time carefully throughout the year. Make sure you are getting enough sleep, that you are getting the right nutrition, and that you are exercising and socialising to maintain a healthy balance so that you don't burn out.

To help you stay on target, plan out a study timetable. One way to do this is to:

1. Assess your current study time and social time. How much are you dedicating to each?
2. List all your commitments and deadlines, including sport, work, assignments, etc.
3. Prioritise the list and reassess your time to ensure you can meet all your commitments.
4. Decide on a format, whether it be weekly or monthly, and schedule in a study routine.
5. Keep your timetable somewhere you can see it.
6. Be consistent.

The exam

The end-of-year examination accounts for 50% of your total mark. It assesses your achievement in the following objectives of units 3 & 4:

1. describe and explain chemical equilibrium systems, oxidation and reduction, properties and structure of organic materials, and chemical synthesis and design.
2. apply understanding of chemical equilibrium systems, oxidation and reduction, properties and structure of organic materials, and chemical synthesis and design.
3. analyse evidence about chemical equilibrium systems, oxidation and reduction, properties and structure of organic materials, and chemical synthesis and design to identify trends, patterns, relationships, limitations or uncertainty.
4. interpret evidence about chemical equilibrium systems, oxidation and reduction, properties and structure of organic materials, and chemical synthesis and design to draw conclusions based on analysis.

The examination includes two papers. Each paper includes a number of possible question types:
- multiple-choice questions
- short-response items requiring single-word, sentence or paragraph responses
- calculating using algorithms
- interpreting graphs, tables or diagrams
- responding to unseen data and/or stimulus.

You will have 90 minutes plus 10 minutes reading time for each paper.

Modified from Chemistry General Senior Syllabus 2019, © State of Queensland (QCAA) 2019, licensed under CC BY 4.0

The day of the exam

The night before your exam, try to get a good rest and avoid cramming, as this will only increase stress levels. On the day of the exam, arrive at the venue of your exam early and bring everything you will need with you. If you have to rush to the exam, you will increase your stress levels, thereby lowering your ability to do well. Further, if you are late, you will have less time to complete the exam, which means that you may not be able to answer all the questions or may rush to finish and make careless mistakes. If you are more than 30 minutes late, you may not be allowed to enter the exam. Don't worry too much about 'exam jitters'. A certain amount of stress is required to help you concentrate and achieve an optimum level of performance. If, however, you're still feeling very nervous, breathe deeply and slowly. Breathe in for a count of six seconds, and out for six seconds until you begin to feel calm.

Perusal time

Use your time wisely! *Do not* use the perusal time to try and figure out the answers to any of the questions until you've read the whole paper! The exam will not ask you a question testing the same knowledge twice, so look for hints in the stem of the question and avoid repeating yourself. Plan your approach so that when you begin writing, you know which section, and ideally which question, you are going to start with.

Strategies for effective responses

Pay particular attention to the cognitive verb used in the question. For example, a question with the cognitive verb 'explain' requires a different response to a question with the cognitive verb 'describe'. Familiarise yourself with the definitions of the commonly used cognitive verbs below (listed in order of complexity). Understanding the definitions of these cognitive verbs will help ensure you are not just providing general information or restating the question without answering it.

describe	give an account (written or spoken) of a situation, event, pattern or process, or of the characteristics or features of something
explain	make an idea or situation plain or clear by describing it in more detail or revealing relevant facts; give an account; provide additional information
apply	use knowledge and understanding in response to a given situation or circumstance; carry out or use a procedure in a given or particular situation
analyse	dissect to ascertain and examine constituent parts and/or their relationships; break down or examine in order to identify the essential elements, features, components or structure; determine the logic and reasonableness of information; examine or consider something in order to explain and interpret it, for the purpose of finding meaning or relationships and identifying patterns, similarities and differences
interpret	use knowledge and understanding to recognise trends and draw conclusions from given information; make clear or explicit; elucidate or understand in a particular way; bring out the meaning of, e.g. a dramatic or musical work, by performance or execution; bring out the meaning of an artwork by artistic representation or performance; give one's own interpretation of; identify or draw meaning from, or give meaning to, information presented in various forms, such as words, symbols, pictures or graphs

Chemistry General Senior Syllabus 2019, © State of Queensland (QCAA) 2019, licensed under CC BY 4.0

Write down all the key steps to show your working when calculating answers. Each mark awarded for a question relates to a step in the calculation. Your *A+ Chemistry Study Notes* book features worked examples and step-by-step worked solutions to help familiarise you with these types of questions.

Common areas of weakness include:

- acid–base nature of indicators
- the relationship between standard reduction potentials and redox reactions
- balancing half-equations in acid conditions
- performing multistep calculations
- reaction pathways.

But be aware that your own weaknesses may be in other areas.

Multiple-choice questions

Read the question carefully and underline any important information to help you break the question down and avoid misreading it. Read all the possible solutions and eliminate any clearly incorrect answers. Fill in the multiple-choice answer sheet carefully and clearly. Check your answer and move on. Do not leave any answers blank.

Short response questions

It is important that you plan your response before writing. To do this, **BUG** the question:

- **B**ox the cognitive verb (describe, explain, apply, etc.).
- **U**nderline any key terms or important information and take note of the mark allocation.
- **G**o back and read the question again.

Many questions require you to apply your knowledge to unfamiliar situations so it is okay if you have never heard of the context before, but you should know which part of the syllabus you are being tested on and what the question is asking you to do. If there is a stimulus included, use information from that as part of the response to show how you are linking the (unfamiliar) context to your knowledge.

Plan your response in a logical sequence. If the question says, 'describe and explain' then structure your answer in that order. You can use dot points to do this, but ensure you write in full sentences. Rote-learned answers are unlikely to receive full marks, so you must relate the concepts of the syllabus back to the question and ensure that you answer the question that is being asked. Planning your response to include the relevant information and the key terminology will help you from writing too much, contradicting yourself, or 'waffling' on and wasting time. If you have time at the end of the paper, go back and re-read your answers.

Good luck. You've got this!

ABOUT THE AUTHOR

Brett Steeples

Brett was a research chemist and lecturer at Manchester University in the UK before becoming a Chemistry teacher. Brett is currently a Lead Confirmer for Chemistry and HOD Science at Ipswich Girls' Grammar School. He was a member of the Chemistry State review panel and has four years' experience writing external Chemistry exam papers. Brett co-authored the *Nelson QCE Chemistry* series.

A+ DIGITAL FLASHCARDS

Revise key terms and concepts online with the A+ Flashcards. Each topic glossary in this book has a corresponding deck of digital flashcards you can use to test your understanding and recall. Just scan the QR code or type the URL into your browser to access them.

https://get.ga/aplus-qce-chem-u34

Note: You will need to create a free NelsonNet account.

UNIT 3
EQUILIBRIUM, ACIDS AND REDOX REACTIONS

Chapter 1
Topic 1: Chemical equilibrium systems

Topic summary

At lower levels of study, chemical reactions are depicted as going to completion. It is understood that the reactants react together to form the products:

$$aA + bB \rightarrow cC + dD$$

This is important when chemical calculations are first encountered that involve amounts of products formed from given amounts of reactants.

However, many reactions do not go to completion and are reversible. This reversibility is indicated in equations by a forward and reverse double arrow:

$$aA + bB \rightleftharpoons cC + dD$$

The introduction of the concept of reversibility has huge implications when calculating amounts of reactants and products. It transforms our understanding of the behaviour of acids and bases, reactions and their real-world applications.

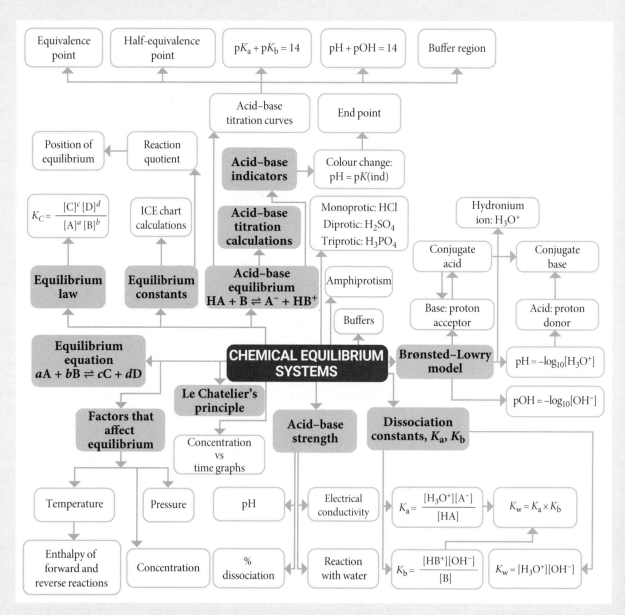

1.1 Chemical equilibrium

1.1.1 Open and closed chemical systems

It is important to understand the difference between an **open system** and a **closed system**.

In an open system, matter and energy can be exchanged with its surroundings.

In a closed system, no substances can enter or leave the system, but energy *can* be exchanged.

Consider the thermal decomposition of calcium carbonate to give calcium oxide and carbon dioxide.

In a closed system, **equilibrium** is established.

$$CaCO_3(s) \rightleftharpoons CaO(s) + CO_2(g)$$

FIGURE 1.1 An example of a closed system

In an open system, the CO_2 gas produced can escape. Equilibrium can never be established.

$$CaCO_3(s) \rightarrow CaO(s) + CO_2(g)$$

FIGURE 1.2 An example of an open system

1.1.2 Physical and chemical changes

Equilibrium systems can involve chemical and/or physical changes.

Physical changes

A **physical change** is one that involves no new products. Consider the changes of state involving water:

$$H_2O(s) \xrightarrow[\text{Freezing}]{\text{Melting}} H_2O(l) \xrightarrow[\text{Condensation}]{\text{Evaporation}} H_2O(g)$$

FIGURE 1.3 The physical changes involving water

These are all physical changes and no new substances are produced, water has simply changed from ice to liquid water to water vapour.

Chemical changes

A **chemical change** is one in which reactants react to produce products that are new substances with different physical and chemical properties.

If magnesium is reacted with hydrochloric acid, a chemical change occurs:

$$Mg(s) + 2HCl\,(aq) \rightarrow MgCl_2(aq) + H_2(g)$$

Magnesium solid has reacted with hydrochloric acid solution to produce new substances, magnesium chloride solution and hydrogen gas.

1.1.3 Equilibrium equations

It is understandable to think of chemical reactions proceeding until one of the reactants (the limiting reagent) is used up. This is called 'going to completion'.

$$aA + bB \rightarrow cC + dD + aA$$

However, many reactions occur when, once products have been formed, they then react together to produce the original reactants. These reactions are **reversible** and are written with the double arrow, $\rightleftharpoons$.

$$aA + bB \rightleftharpoons cC + dD$$

The $\rightharpoonup$ and the $\leftharpoonup$ half-arrows represent the relative rates of the two reactions:

$$aA + bB \rightharpoonup cC + dD$$

This is referred to as the FORWARD reaction. A and B are the reactants, C and D are the products.

$$aA + bB \leftharpoonup cC + dD$$

This is referred to as the REVERSE reaction. C and D are the reactants, A and B are the products. When the rates of the forward and reverse reactions are the same, the reaction is at equilibrium.

1.1.4 Dynamic equilibrium

In a closed system involving a reversible reaction, reactants form products and products form reactants. These forward and reverse reactions are occurring constantly, but the *overall* system does not change. The system is said to be in a state of **dynamic equilibrium**.

For an equilibrium system:

- the rate of the forward reaction will decrease as the concentration of the reactants decreases.

- the rate of the reverse reaction will increase as the concentration of the products increases.

This continues until the rates of the forward and reverse reactions are the same.

1.1.5 Reversibility

The rate of a chemical reaction is governed by its activation energy, E_a. If the E_a of a reaction is very high, it is unlikely to occur.

Consider the reaction:

Reactants $\rightleftharpoons$ Products

The E_a of the forward and reverse reactions can be deduced from their energy profile diagrams:

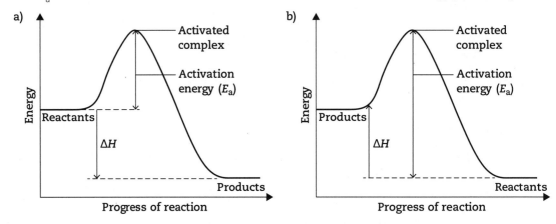

FIGURE 1.4 Energy profile diagrams showing activation energy for a) forward and b) reverse reactions

From these diagrams, it is clear that the activation energy for the forward reaction is much less than that for the reverse reaction. Therefore, the forward reaction is more likely to occur.

1.1.6 The position of equilibrium

It is possible to use graphical methods to illustrate the behaviour of chemical equilibrium systems and determine the position of equilibrium at any given time. The two most common graphical methods are:

- reaction rate against time

- concentration against time.

Reaction rate versus time

Figure 1.5 shows the rate versus time graph for the reaction:

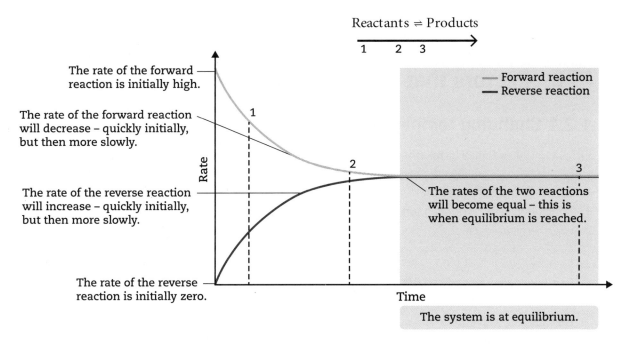

FIGURE 1.5 Reaction rate vs time graph for the reaction: Reactants ⇌ Products

Point **1** on the graph corresponds to an equilibrium position that is far to the left. The rate of the forward reaction is much greater than that of the reverse reaction.

Point **2** corresponds to a position close to equilibrium, the rates of the forward and reverse reactions are similar.

Point **3**, the system is at equilibrium. The lines are horizontal, there is no change in the reaction rate.

Concentration versus time

Figure 1.6 shows the concentration versus time graph for the reaction:

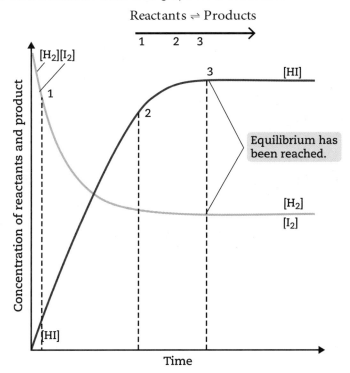

FIGURE 1.6 Concentration vs time graph for the reaction: Reactants ⇌ Products

At point **1**, the concentration of reactants is high compared to the products.

At point **2**, the concentration of reactants is low compared to the products.

At point **3**, the concentrations of reactants and products are steady and parallel – they do not change with time. The system is at equilibrium.

1.2 Factors that affect equilibrium

1.2.1 Changing temperature

When dealing with how temperature affects equilibrium systems, a reaction with a measurable enthalpy change must be considered. This is one that is represented by a thermochemical equation. A **thermochemical equation** is simply a balanced equation with its corresponding enthalpy changed.

For example, the production of ammonia from nitrogen and hydrogen can be represented by the thermochemical equation:

$$N_2(g) + 3H_2(g) \rightleftharpoons 2NH_3(g) \quad \Delta H = -92.4 \text{ kJ}$$

Reading the equation from left to right, the enthalpy is -92.4 kJ, which means that the production of $NH_3(g)$ is an exothermic process. It releases heat to its surroundings.

BUT as this is an equilibrium system, the reverse reaction must be considered:

Forward reaction → ΔH = – (negative) → exothermic → produces heat

$$N_2(g) + 3H_2(g) \rightarrow 2NH_3(g) \quad \Delta H = -92.4\,\text{kJ}$$

Reverse reaction → ΔH = + (positive) → endothermic → absorbs heat

$$N_2(g) + 3H_2(g) \leftarrow 2NH_3(g) \quad \Delta H = 92.4\,\text{kJ}$$

Increasing the temperature

If heat is added to this system, the reaction will work to counteract the change by shifting in the direction that will absorb the added heat.

It can do this by increasing the rate of the reverse reaction:

$$N_2(g) + 3H_2(g) \rightleftharpoons 2NH_3(g) \quad \Delta H = -92.4 \text{ kJ}$$

Remember, the reverse reaction is endothermic and absorbs heat.

This process is known as *shifting to the left*.

Decreasing the temperature

If heat is removed from this system, the reaction will work to counteract the change by shifting in the direction that will release heat to replace the heat that was removed.

It can do this by increasing the rate of the forward reaction:

$$N_2(g) + 3H_2(g) \rightleftharpoons 2NH_3(g) \quad \Delta H = -92.4 \text{ kJ}$$

Remember, the forward reaction is exothermic and releases heat.

This process is known as *shifting to the right*.

These changes can be tracked by referring to a concentration versus time graph, as shown in Figure 1.7.

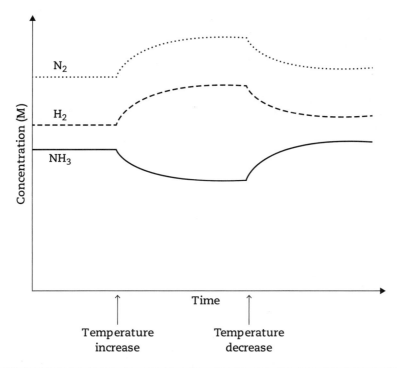

FIGURE 1.7 Concentration vs time graph showing temperature changes for the reaction: $N_2(g) + 3H_2(g) \rightleftharpoons 2NH_3(g)$ $\Delta H = -92.4$ kJ

> **Hint**
> Changes in temperature are easy to spot on these graphs. Everything changes gradually at the same time. This is because heating and cooling are NOT instantaneous processes.

1.2.2 Changing concentration

A change in the concentration of a substance in an equilibrium system has implications for the whole system.

Consider the reaction:

$$2SO_2(g) + O_2(g) \rightleftharpoons 2SO_3(g)$$

If, for example, some $SO_3(g)$ was removed from the system (that is, its concentration was decreased), then the rate of the forward reaction would increase to replace the SO_3 that was lost:

$$2SO_2(g) + O_2(g) \rightleftharpoons 2SO_3(g)$$

But if the rate of the forward reaction is increased, then the relative amounts of the reactants, $SO_2(g)$ and $O_2(g)$, will decrease a little because they are being used up more quickly.

If some $O_2(g)$ was added to the system, then the rate of the forward reaction would increase to remove the excess $O_2(g)$.

In this case, the amount of $SO_2(g)$ would decrease a little and the amount of $SO_3(g)$ would increase a little.

These changes can be tracked by referring to a concentration versus time graph, as shown in Figure 1.8.

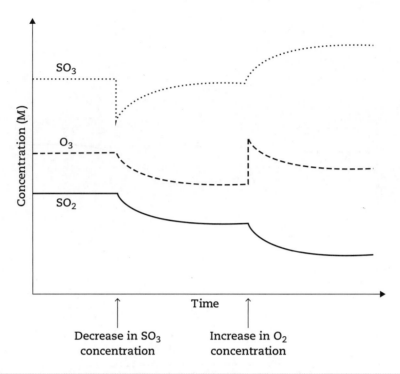

FIGURE 1.8 Concentration vs time graph showing concentration changes for the reaction: $2SO_2(g) + O_2(g) \rightleftharpoons 2SO_3(g)$

1.2.3 Changing volume/pressure

Changing the volume of a gaseous equilibrium system affects the pressure of the system. This is illustrated in Figures 1.9 and 1.10.

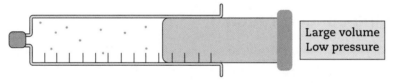

FIGURE 1.9 A gaseous system at equilibrium in a gas syringe

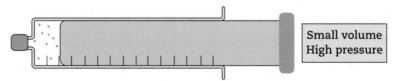

FIGURE 1.10 A gaseous system under pressure

It is clear in Figure 1.10 that the plunger of the gas syringe has been pushed further into the syringe. This has decreased the volume but has increased the pressure.

When discussing the effects of changes in pressure on gaseous equilibrium systems, it is important to examine the **stoichiometry** of the reaction involved.

Consider the production of ammonia:

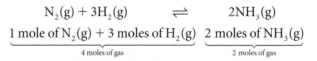

The equation has 4 moles of gas on the left-hand side and 2 moles of gas on the right-hand side. The more moles of gas there are, the higher the pressure.

If the overall pressure on this reaction was increased, the system would shift to the right-hand side (the lower pressure side) to counteract the increase in pressure. If the overall pressure was decreased, the system would shift to the left to increase the pressure.

These changes can be tracked by referring to a concentration versus time graph, as shown in Figure 1.11.

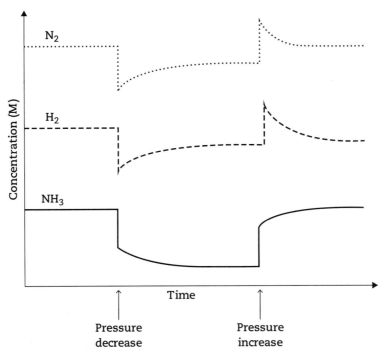

FIGURE 1.11 Concentration vs time graph showing pressure changes for the reaction: $N_2(g) + 3H_2(g) \rightleftharpoons 2NH_3(g)$

It should be noted that for a reaction in which there are equal numbers of molecules of gas on both sides of the equation, such as $H_2(g) + Cl_2(g) \rightleftharpoons 2HCl(g)$, changes in pressure will have no effect on the position of equilibrium.

1.2.4 Adding a catalyst

Catalysts generally increase the rate of reaction. In an equilibrium system, a catalyst will increase the rate of the forward and reverse reactions equally. Therefore, adding a catalyst to an equilibrium system does not affect the equilibrium, it simply speeds up the rate at which equilibrium is established.

1.2.5 Le Chatelier's principle

In previous sections of this chapter, whenever a change to an equilibrium system has been discussed, mention has been made of the system responding in a certain way.

The French chemist, Henri Louis Le Chatelier summarised the effects of changes to equilibrium systems into one law or principle.

Hint

Key concept

'If a system at equilibrium is subject to a change in conditions, then the system will behave in such a way as to partially counteract the change.'

Consider the reaction:

$$2SO_2(g) + O_2(g) \rightleftharpoons 2SO_3(g) \quad \Delta H = -198\,\text{kJ}$$

Le Chatelier's principle can be used to predict the system's response to any changes made to it.

Adding a reactant such as $SO_2(g)$	
	Shifts *right* to counteract the change. This increases the rate of the forward reaction, which partially uses up the extra $SO_2(g)$.
Removing a reactant such as $SO_2(g)$	
Shifts *left* to counteract the change. This increases the rate of the reverse reaction, which produces more $SO_2(g)$ to partially replace that which was removed.	
Adding a product such as $SO_3(g)$	
Shifts *left* to counteract the change. This increases the rate of the reverse reaction, which uses up the extra $SO_3(g)$.	
Heating the system	
Shifts *left* to counteract the change. The forward reaction is exothermic, the reverse reaction is endothermic. If heat is added, the rate of the reverse reaction is increased, which absorbs some of the heat that was added.	
Cooling the system	
	Shifts *right* to counteract the change. If heat is removed, the rate of the forward reaction is increased, which produces heat to replace some of the heat removed.
Increasing the pressure	
	Shifts *right* to counteract the change. The right-hand side has fewer gas molecules and so is the lower pressure side. The rate of the forward reaction is increased, which lowers the pressure.

1.3 Equilibrium constants

1.3.1 The equilibrium law

Consider the equilibrium equation:

$$aA + bB \rightleftharpoons cC + dD$$

The **equilibrium expression** for this can be written as:

$$K_C = \frac{[C]^c[D]^d}{[A]^a[B]^b}$$

As an example, the equilibrium expression for the reaction $N_2(g) + 3H_2(g) \rightleftharpoons 2NH_3(g)$ is:

Reactants   Products
$$N_2(g) + 3H_2(g) \rightleftharpoons 2NH_3(g)$$

Coefficients become powers

$$K_c = \frac{Products}{Reactants} = \frac{[NH_3]^2}{[N_2][H_2]^3}$$

The equilibrium constant with respect to…   Concentration   Square brackets, [] indicate concentration

This is for a **homogeneous system**, in which all of the substances are in the same phase, in this case, gas.

Heterogeneous systems can be more complex.

Consider the reaction:

$$C(s) + H_2O(g) \rightleftharpoons H_2(g) + CO(g)$$

The equilibrium expression for this reaction is:

$$K_c = \frac{[H_2][CO]}{[H_2O]}$$

> **Hint**
> Solids (s) and liquids (l) NEVER appear in an equilibrium expression.

1.3.2 K_c values and the position of equilibrium

The magnitude of an equilibrium constant can provide valuable information about equilibrium systems.

> **Hint**
> **Key concept**
> The equilibrium position is the point in a chemical reaction at which the concentrations of reactants and products do not change.

For any given equilibrium system, such as:

Reactants $\rightleftharpoons$ Products

A small value of K_c means that the reactants' concentrations are higher than the products. The equilibrium lies to the left.

If K_c is close to 1, the reactant and product concentrations are about the same. Equilibrium lies in the centre.

A large value of K_c means that the products' concentrations are higher than the reactants. The equilibrium lies to the right.

1.3.3 The reaction quotient

The ratio of products to reactants only equals the equilibrium constant (K_c) when the system is at equilibrium.

If the system is not at equilibrium this ratio can be calculated and compared to the equilibrium constant to provide information on the extent of the equilibrium.

This ratio is called the **reaction quotient, Q**.

Consider the system:

$$PCl_3(g) + Cl_2(g) \rightleftharpoons PCl_5(g)$$

If, a certain time after mixing, the concentration of all the substances in the system was measured at 25°C and found to be:

$$[PCl_3] = [Cl_2] = 0.125 \, mol \, L^{-1}, [PCl_5] = 2.3 \, mol \, L^{-1}$$

is the system at equilibrium given that K_c at this temperature is 0.18?

$$Q = \frac{[PCl_5]}{[PCl_3][Cl_2]} = \frac{2.3}{0.125 \times 0.125} = 147.2$$

Q is much greater than K_c. The system is not at equilibrium.

So, when comparing Q and K_c values, if:

$Q < K_c$, the system lies to the left and needs to shift right to reach equilibrium.

$Q = K_c$, the system is at equilibrium.

$Q > K_c$, the system lies to the right and must shift left to reach equilibrium.

1.3.4 Calculating equilibrium constants and equilibrium concentrations

Two main types of calculations are encountered here.

Calculating K_c from equilibrium concentrations

Worked example

Question: Calculate the equilibrium constant for the system at 350°C.

$$2NOCl(g) \rightleftharpoons 2NO(g) + Cl_2(g)$$

The concentrations of the substances at equilibrium are:

$[NOCl] = 1.5 \, mol \, L^{-1}, [NO] = [Cl_2] = 0.4 \, mol \, L^{-1}$

> **Hint**
>
> In calculation questions, the units of equilibrium constants are NOT required.

$$Q = \frac{[NO]^2[Cl_2]}{[NOCl]^2} = \frac{0.4^2 \times 0.4}{1.5^2} = 0.028$$

Calculating K_c from initial concentrations

Worked example

Question: A 0.25 mol sample of carbonyl fluoride gas, COF_2, was placed in a 1 L container at 300°C and allowed to come to equilibrium according to the following equation:

$$COF_2(g) \rightleftharpoons CO(g) + F_2(g)$$

If the equilibrium constant, K_c, for this reaction is 1.8×10^{-4}, calculate the equilibrium concentration of all the species present at equilibrium.

Step 1

Set up an **ICE** chart.

	COF$_2$	CO	F$_2$
Initial	0.25 M	0 M	0 M
Change	−x	+x	+x
Equilibrium	0.25 − x	x	x

Step 2

Substitute the **E**quilibrium amounts into the equilibrium expression along with the value for the equilibrium constant and calculate the value of x.

$$K_c = \frac{[CO][F_2]}{[COF_2]} = 1.8 \times 10^{-4} = \frac{x^2}{0.25 - x}$$

$$x = 6.7 \times 10^{-3}$$

Hint

In these calculations, it is assumed that the value of x is very small, therefore it is assumed that initial $-x$ is approximately equal to initial. You MUST mention this assumption in an answer.

Step 3

Finish the question by substituting the value of x into the **E**quilibrium line of the ICE chart.

$$[COF_2] = 0.25 - 6.7 \times 10^{-3} = 0.2433 \text{ M}$$

$$[CO] = 6.7 \times 10^{-3} \text{ M}$$

$$[F_2] = 6.7 \times 10^{-3} \text{ M}$$

Worked example

Question: The equilibrium constant for the Haber process is 0.62. An experiment is carried out in which 3.5 moles of $N_2(g)$ and 12.0 moles of $H_2(g)$ are placed in a 2 L container and allowed to reach equilibrium according to the equation:

$$N_2(g) + 3H_2(g) \rightleftharpoons 2NH_3(g)$$

If 5.8 moles of ammonia is present at equilibrium, what is the concentration of hydrogen gas at equilibrium?

Step 1

Calculate the molar concentrations of all species given that the volume of the container is 2 L.

$$[N_2] = \frac{3.5}{2} = 1.75 \text{ M}$$

$$[H_2] = \frac{12.0}{2} = 6.00 \text{ M}$$

$$[NH_3] = \frac{5.8}{2} = 2.9 \text{ M}$$

Step 2

Set up an **ICE** chart.

	N$_2$	H$_2$	NH$_3$
Initial	1.8 M	6.00 M	0 M
Change	$-x$	$-3x$	$+2x$
Equilibrium	$1.8 - x$	$6.00 - 3x$	$2x$

$$N_2(g) + 3H_2(g) \rightleftharpoons 2NH_3(g)$$

These numbers come directly from the coefficients in the balanced chemical equation.

Step 3

Using the equilibrium concentration of ammonia given, determine the value of x.

From the ICE chart, the equilibrium $[NH_3] = 2x$

BUT

From the question:

$[NH_3] = 2.9$ M

Therefore:

$$[NH_3] = 2.9 \text{ M} = 2x, \frac{2.9}{2} = 1.45 \text{ M}$$

Step 4

Substitute the value of x into the equilibrium line for H_2.

$$[H_2] = 6.00 - (3 \text{ multiplied by } 1.45) = \mathbf{1.65\,M}$$

1.4 Properties of acids and bases

1.4.1 What is an acid?

When the concept of acids is first introduced in your Chemistry studies, it is defined using the Arrhenius theory that states an acid is a substance that produces hydrogen ions, H^+, in water:

$$HCl(aq) \rightarrow H^+(aq) + Cl^-(aq)$$

This does not tell the whole story. The H^+ ion is a hydrogen atom without its electron. It is a proton and, as such, is a highly concentrated positive charge. As soon as it is produced by the acid, it instantly attaches itself to a nearby water molecule, to form the hydronium ion, $H_3O^+(aq)$.

The equation above becomes:

$$HCl(aq) + H_2O(aq) \rightarrow H_3O^+(aq) + Cl^-(aq)$$

Acids can be classified by the number of H^+ ions (protons) they release in water.

Hydrochloric acid (HCl, shown in the above equation) releases one $H^+(aq)$ ion and so is termed **monoprotic**.

Polyprotic acids are acids that produce more than one H^+ ion in solution.

For example, sulfuric acid dissociates (breaks down) completely via two stages to produce two H^+ ions in solution and therefore, two hydronium ions. It is polyprotic and, more specifically, **diprotic**.

$$H_2SO_4(l) + H_2O(l) \rightarrow \mathbf{HSO_4^-}(aq) + H_3O^+(aq)$$
$$\text{hydrogen sulfate ion}$$

$$HSO_4^-(aq) + H_2O(l) \rightarrow \mathbf{SO_4^{2-}}(aq) + H_3O^+(aq)$$
$$\text{sulfate ion}$$

Phosphoric acid dissociates completely via three stages to produce three hydronium ions. It is polyprotic and, more specifically, **triprotic**.

$$H_3PO_4(s) + H_2O(l) \rightarrow \mathbf{H_2PO_4^-}(aq) + H_3O^+(aq)$$
$$\text{dihydrogen phosphate ion}$$

$$\mathbf{H_2PO_4^-}(aq) + H_2O(l) \rightarrow \mathbf{HPO_4^{2-}}(aq) + H_3O^+(aq)$$
$$\text{hydrogen phosphate ion}$$

$$\mathbf{HPO_4^{2-}}(aq) + H_2O(l) \rightarrow \mathbf{PO_4^{3-}}(aq) + H_3O^+(aq)$$
$$\text{phosphate ion}$$

1.4.2 Acid and base strength

The strength of an acid or a base is a measure of how readily it dissociates in water.

It is useful to think of the dissociation of an acid or a base as an equilibrium system in which water is an integral part of the system:

$$HCl(aq) + H_2O(l) \rightleftharpoons H_3O^+(aq) + Cl^-(aq)$$

BUT:

HCl is a strong acid, which means that all the HCl molecules dissociate. The equilibrium system can be rewritten slightly:

$$\underset{0\%}{HCl(aq) + H_2O(l)} \rightleftharpoons \underset{100\%}{H_3O^+(aq) + Cl^-(aq)}$$

In this system, the equilibrium lies far to the right.

Ethanoic acid (CH_3COOH) is a weak acid, which means that only about 1% of the acid molecules dissociate:

$$\underset{\approx 99\%}{CH_3COOH(aq) + H_2O(l)} \rightleftharpoons \underset{\approx 1\%}{H_3O^+(aq) + CH_3COO^-(aq)}$$

In this system, the equilibrium lies far to the left.

A base produces hydroxide ions, OH^-, in solution.

A typical strong base is sodium hydroxide:

$$\underset{0\%}{NaOH(s) + H_2O(l)} \rightleftharpoons \underset{100\%}{Na^+(aq) + H_2O(l) + OH^-(aq)}$$

A typical weak base is ammonia, NH_3:

$$\underset{\approx 99\%}{NH_3(aq) + H_2O(l)} \rightleftharpoons \underset{\approx 1\%}{NH4^+(aq) + OH^-(aq)}$$

Hint

Acid strength and molar concentration should not be confused.

Figure 1.12 shows the difference between strong, weak, concentrated and dilute acids.

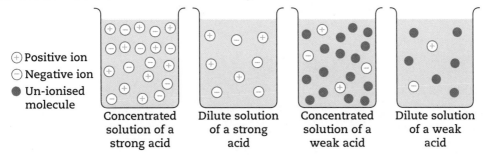

⊕ Positive ion
⊖ Negative ion
● Un-ionised molecule

Concentrated solution of a strong acid

Dilute solution of a strong acid

Concentrated solution of a weak acid

Dilute solution of a weak acid

FIGURE 1.12 Concentrated and dilute solutions of a strong acid and a weak acid

Acid–base strength and electrical conductivity

From Figure 1.12, it is clear that strong acids produce more positive and negative ions in solution than weak ones.

It would be easy, therefore, to distinguish strong and weak acids of the same concentration by measuring their respective electrical conductivities, as shown in Figure 1.13.

Strong acid = more ions = higher electrical conductivity.

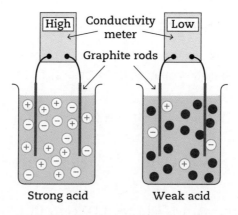

FIGURE 1.13 Apparatus for measuring the conductivity of strong and weak acids

1.5 | pH scale

1.5.1 Self-ionisation of water

It is important to note that in pure water, some water molecules react with each other:

$$H_2O(l) + H_2O(l) \rightleftharpoons H_3O^+(aq) + OH^-(aq)$$

The equilibrium constant for this system is very small and is referred to as the **self-ionisation constant, K_w,** or the ionic product of water. The equilibrium expression for this system can be written:

$$K_w = [H^+][OH^-], \text{ where } K_w = 10^{-14} M^2 \text{ at } 25°C$$

This is extremely important because it enables the concentrations of H^+ ions and OH^- ions in acidic and basic solutions to be determined.

Worked example

Question: Calculate the concentrations of H^+ ions and OH^- ions in a 1.85 M solution of HNO_3.

Step 1

Write the dissociation equation.

$$HNO_3(aq) + H_2O(l) \rightarrow H_3O^+(aq) + NO_3^-(aq)$$

Remember that nitric acid is a strong acid, meaning that *every* HNO_3 molecule dissociates.

> **Hint**
>
> For the purposes of this course, the only strong acids required are hydrochloric acid (HCl), nitric acid (HNO_3) and sulfuric acid (H_2SO_4).

Step 2

Determine $[H^+]$.

HNO_3 is a strong, monoprotic acid. So:

$$1.85\,M\,HNO_3 \rightarrow \mathbf{1.85\,M\,H^+}$$

Step 3

Using K_w, determine the $[OH^-]$.

$$K_w = [H^+][OH^-] = 10^{-14}$$

$$1.85 \times [OH^-] = 1 \times 10^{-14}$$

$$[OH^-] = \frac{1 \times 10^{-14}}{1.85}$$

$$[OH^-] = \mathbf{5.4 \times 10^{-15}\,M}$$

This very small value is understandable – HNO_3 is a strong acid!

Worked example

Question: Calculate the concentrations of H^+ ions and OH^- ions in a 1.85 M solution of $Ba(OH)_2$.

Step 1

Write the dissociation equation.

$$Ba(OH)_2(aq) + 2H_2O(l) \rightarrow Ba^{2+}(aq) + 2H_2O(l) + 2OH^-(aq)$$

Remember that barium hydroxide is a strong base, meaning *every* $Ba(OH)_2$ molecule dissociates. So, for every $Ba(OH)_2$ that dissociates, *two* OH^- ions are produced.

Step 2

Determine $[OH^-]$.

$$1.85\,M\ Ba(OH)_2 \rightleftharpoons 1.85 \times 2 = \mathbf{3.7\,M\ OH^-}$$

Step 3

Using K_w, determine the $[H^+]$.

$$3.7 \times [H^+] = 1 \times 10^{-14}$$

$$[H^+] = \frac{1 \times 10^{-14}}{3.7}$$

$$[H^+] = \mathbf{2.7 \times 10^{-15}\,M}$$

This very small value is understandable – $Ba(OH)_2$ is a strong base!

Hint

Strong bases include most metal hydroxides, particularly groups I and II, and groups I and II oxides.

1.5.2 The pH scale equation

The pH scale is a measure of the concentration of hydronium ions and is given by the equation:

$$pH = -\log_{10}[H_3O^+]$$

Worked example

Question: Calculate the pH of a 0.38 M HCl solution.

HCl is a strong acid so 0.38 M HCl = $[H_3O^+]$:

$$pH = -\log_{10} 0.38 = \mathbf{0.42}$$

Hint

pH is a scale without units.

Worked example

Question: Calculate the pH of a 0.38 M H_2SO_4 solution.

H_2SO_4 is a strong, diprotic acid; for every molecule of H_2SO_4 that dissociates, two hydronium ions are produced.

$$0.38\,M\ H_2SO_4 \times 2 = 0.76\,M = [H_3O^+]$$

$$pH = -\log_{10} 0.76 = \mathbf{0.12}$$

Worked example

Question: Calculate the pH of a 0.8 M NaOH solution.

Step 1

Determine $[OH^-]$.

$$0.8\,M\ NaOH = 0.8\,M\ [OH^-]$$

Step 2

Using the K_w equation, determine the $[H^+]$.

$$[OH^-][H^+] = 1.00 \times 10^{-14}$$

$$0.8 \times [H^+] = 1 \times 10^{-14}$$

$$[H^+] = \frac{1 \times 10^{-14}}{3.8}$$

$$[H^+] = 1.2 \times 10^{-14}\,M$$

Step 3

Calculate pH:

$$pH = -\log[H^+]$$

$$pH = -\log_{10} 1.2 \times 10^{-14} = \mathbf{13.9}$$

The pH scale runs from 1 through to 14, as shown in Figure 1.14.

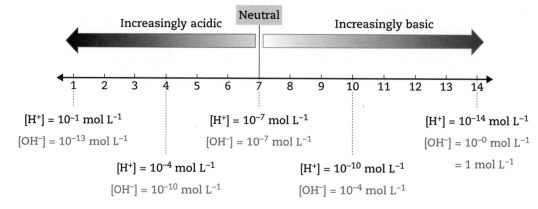

FIGURE 1.14 The pH scale

The pOH scale

Just as the pH scale is a measure of the concentration of hydronium ions, the pOH scale is a measure of the concentration of hydroxide ions.

The pOH of a substance is given by:

$$pOH = -\log_{10}[OH^-]$$

Worked example

Question: Calculate the pOH of a 0.14 M KOH solution.

$$pOH = -\log_{10} 0.14 = \mathbf{0.85}$$

Determining the resulting pH from an acid–base reaction

Worked example

Question: Ethanoic acid reacts with NaOH according to the equation:

$$CH_3COOH(aq) + NaOH(aq) \rightleftharpoons CH_3COO^-(aq) + Na^+ + H_2O(l)$$

Calculate the pH of the solution produced by mixing 253.0 mL of a 0.15 M solution of ethanoic acid with 262.0 mL of a 0.08 M sodium hydroxide solution.

(K_a ethanoic acid = 1.8×10^{-5})

Step 1

Draw a diagram of the situation.

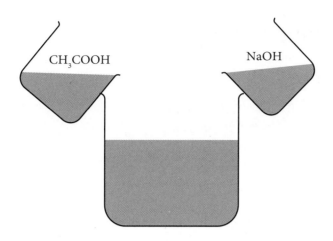

Step 2

Calculate the number of moles of ethanoic acid added, ensuring you convert mL to L in your calculations.

$$c = 0.15 \text{ M}$$
$$V = 253.0 \text{ mL}$$
$$n = cV$$
$$n = 0.15 \times \frac{253.0}{1000}$$
$$n = 0.038 \text{ mol}$$

Step 3

Calculate the number of moles of NaOH added, ensuring you convert mL to L in your calculations.

$$c = 0.08 \text{ M}$$
$$V = 262.0 \text{ mL}$$
$$n = cV$$
$$n = 0.08 \times \frac{262.0}{1000}$$
$$n = 0.021 \text{ mol}$$

Step 4

Calculate the number of moles of acid (or base) remaining after neutralisation.

The number of moles of ethanoic acid is greater than NaOH so, given that they react in a 1:1 ratio, there will be acid left over.

0.038 − 0.021 = 0.017 moles ethanoic acid remaining unreacted.

Step 5

Calculate the concentration of ethanoic acid remaining, ensuring you convert from mL to L in the denominator.

$$c = \frac{c}{V} = \frac{\left(\dfrac{0.017}{253.0 + 262.0} \right)}{1000}$$
$$c = 0.033 \text{ M}$$

Step 6

Calculate the pH of the 0.033 M ethanoic acid solution given that its $K_a = 1.8 \times 10^{-5}$.

$$1.8 \times 10^{-5} = \frac{x^2}{0.033}$$
$$x = [H_3O^+] = 5.95 \times 10^{-7}$$
$$pH = -\log_{10} 5.59 \times 10^{-7} = 6.2$$

1.6   Brønsted–Lowry model

1.6.1 Conjugate acid–base pairs

This is a much more inclusive model of acids and bases than the Arrhenius theory.

According to the Brønsted–Lowry model:

- an acid is a substance that donates one or more protons or hydrogen ions (H^+).

- a base is redefined as a substance that accepts one or more protons.

- using this model, it is possible to describe acids and bases in terms of equilibrium systems:

$$HA(aq) + H_2O(l) \rightleftharpoons H_3O^+(aq) + A^-(aq)$$

When an acid donates its proton to a base, what is left behind is called the **conjugate base**. When a base accepts a proton from an acid, the product is called the **conjugate acid**. This relationship is shown in Figure 1.15.

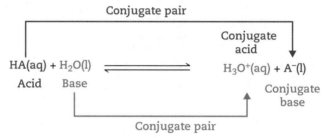

FIGURE 1.15 The relationship between conjugate acid–base pairs

1.6.2 Amphiprotic substances

Amphiprotic substances can act as acids or bases.

Consider the dihydrogen phosphate ion, $H_2PO_4^-$.

Acting as an acid, it gives away its H^+:

$$H_2PO_4^-(aq) + H_2O(l) \rightleftharpoons HPO_4^{2-}(aq) + H_3O^+(aq)$$
$$\text{acid} \qquad \text{base} \qquad \text{conjugate base} \quad \text{conjugate acid}$$

Acting as a base, an H^+ is added:

$$H_2PO_4^-(aq) + H_2O(l) \rightleftharpoons H_3PO_4(aq) + OH^-(aq)$$
$$\text{base} \qquad \text{acid} \qquad \text{conjugate acid} \quad \text{conjugate base}$$

1.6.3 Buffers

Buffers are solutions that can maintain a pH within a certain range, despite an acid or a base being added.

There are two main types of buffer solutions:

Acidic buffer solutions

These consist of weak acid in equilibrium with its conjugate base; for example, the ethanoic acid/ethanoate ion buffer solution:

$$CH_3COOH(aq) + H_2O(l) \rightleftharpoons CH_3COO^-(aq) + H_3O^+(aq)$$
$$\text{ethanoic acid} \qquad\qquad \text{ethanoate ion}$$
$$\text{weak acid} \qquad\qquad\qquad \text{conjugate base}$$

9780170459150

If a small amount of acid (H_3O^+) is added:

- initially, the pH decreases slightly because of the increased number of H_3O^+ ions present
- then, the conjugate base CH_3COO^- ions react with the excess H_3O^+. The equilibrium is restored, and the pH increases to close to its original value.

If a small amount of base (OH^-) is added:

- initially, the pH increases slightly because of the increased number of OH^- ions present
- then, CH_3COOH dissociates, producing more H_3O^+ ions, which neutralise the excess OH^- ions. The equilibrium is restored and the pH decreases to close to its original value.

Basic buffer solutions

These consist of weak acid in equilibrium with its conjugate base. For example, the ammonia/ammonium ion buffer solution:

$$NH_3(aq) + H_2O(l) \rightleftharpoons NH_4^+(aq) + OH^-(aq)$$

ammonia	ammonium ion
weak base	conjugate acid

If a small amount of acid is added:

- initially, the pH decreases because of the increased number of H_3O^+ ions
- then, more NH_3 reacts, producing more OH^- ions that neutralise the excess H_3O^+ ions. The equilibrium is restored, and the pH increases to close to its original value.

If a small amount of base (OH^-) is added:

- initially, the pH increases slightly because of the increased number of OH^- ions present
- then, the conjugate acid (NH_4^+) reacts with the excess OH^- ions. The equilibrium is restored, and the pH decreases to close to its original value.

1.7 Dissociation constants

1.7.1 Dissociation constants and acid strength

Since most acids are weak, their dissociation in water can be represented as an equilibrium system such as:

$$CH_3COOH(aq) + H_2O(l) \rightleftharpoons H_3O^+(aq) + CH_3COO^-(aq)$$

This, in turn, can be represented by an equilibrium expression:

$$K = \frac{[H_3O^+][CH_3COO^-]}{[CH_3COOH]}$$

Because it is referring to the dissociation of an acid, the equilibrium constant is represented by K_a. The K_a value of an acid is a measure of the strength of the acid.
The bigger the K_a, the stronger the acid.

Dissociation constants for bases

$$NH_3(aq) + H_2O(l) \rightleftharpoons NH_4^+(aq) + OH^-(aq)$$

$$K_b = \frac{[NH_4^+][OH^-]}{[NH_3]}$$

1.7.2 Calculations involving dissociation constants

There are several different types of calculations involving dissociation constants.

Calculating K_a from a given pH

Worked example

Question: A 0.14 M solution of hydrocyanic acid, HCN, has a pH of 4.5. Calculate the K_a value for this acid.

Step 1

Write the dissociation equation.

$$HCN(aq) + H_2O(l) \rightleftharpoons H_3O^+(aq) + CN^-(aq)$$

Step 2

Write the equilibrium expression.

$$K_a = \frac{[H_3O^+][CN^-]}{[HCN]}$$

Step 3

Use the pH value to calculate the $[H_3O^+]$.

$$4.5 = -\log_{10}[H_3O^+]$$

$$10^{-4.5} = [H_3O]$$

$$[H_3O^+] = 3.2 \times 10^{-5}\,M$$

Step 4

Substitute the value for $[H_3O^+]$ and $[CN^-]$ into the equilibrium expression and calculate K_a.

$$K_a = \frac{3.2 \times 10^{-5} \times 3.2 \times 10^{-5}}{0.14}$$

$$K_a = 7.3 \times 10^{-9}$$

> **Hint**
> It is assumed that, because the acid is weak, the amount lost will be negligible so that its initial concentration does not change noticeably.

Calculating pH from a given K_a

Worked example

Question: Determine the pH of a 0.1 M solution of ethanoic acid ($K_a = 1.8 \times 10^{-5}$).

Step 1

Write the dissociation equation.

$$CH_3COOH(aq) + H_2O(l) \rightleftharpoons H_3O^+(aq) + CH_3COO^-(aq)$$

Step 2

Write the equilibrium expression, remembering that $[H_3O^+] = [CH_3COO^-]$.

$$K_a = \frac{[H_3O^+][CH_3COO^-]}{[CH_3COOH]} = \frac{[H_3O^+]^2}{[CH_3COOH]}$$

Step 3

Substitute values into the equation.

$$1.8 \times 10^{-5} = \frac{[H_3O^+]^2}{0.1}$$
$$[H_3O^+] = \sqrt{(1.8 \times 10^{-5} \times 0.1)}$$
$$[H_3O^+] = 1.34 \times 10^{-3}\ M$$

Step 4

Calculate pH.

$$pH = -\log_{10} 1.34 \times 10^{-3}$$
$$pH = \mathbf{2.9}$$

Determining percentage ionisation

Worked example

Question: Determine the percentage ionisation of hypochlorous acid, HClO, in 0.2 M solution that has a pH of 3.1.

> **Hint**
>
> **Key formula**
>
> $$\%\ ionisation = \frac{[A^-]}{[HA]} \times 100$$

Step 1

Write the dissociation equation.

$$HClO(aq) + H_2O(l) \rightleftharpoons H_3O^+(aq) + ClO^-(aq)$$

Step 2

Use the pH value to calculate the $[H_3O^+]$ and, therefore, $[ClO^-]$.

$$3.1 = -\log_{10} [H_3O^+]$$

$$10^{-3.1} = [H_3O]$$

$$[H_3O^+] = [ClO^-] = 7.9 \times 10^{-4}\ M$$

Step 3

Substitute into the formula to determine percentage ionisation of hypochlorous acid.

$$\%\ ionisation = \frac{[A^-]}{[HA]} = \frac{[ClO^-]}{[HClO]} = \frac{[7.9 \times 10^{-4}]}{[0.2]} \times 100 = 0.4\%$$

Summary of the important relationships involving acids and bases

$$pH = -\log_{10} [H_3O^+] \qquad K_w = [H_3O^+][OH^-] \qquad K_a = \frac{[H_3O^+][A^-]}{[HA]}$$

$$pOH = -\log_{10} [OH^-] \qquad K_w = K_a \times K_b$$

$$pH + pOH = 14 \qquad\qquad\qquad K_b = \frac{[BH^+][OH^-]}{[B]}$$

> **Hint**
>
> The formulas above, with the following two exceptions, appear on page 1 of the QCAA *Chemistry formula and data booklet*, which you will have access to in your exam. The two exceptions that do not appear in the *Chemistry formula and data booklet* are:
>
> $$\%\ ionisation = \frac{[A^-]}{[HA]}$$
>
> $$pH + pOH = 14$$

1.8 Acid–base indicators

1.8.1 Components of acid–base indicators

Acid–base indicators are substances that change colour depending on the pH of the solution they are in.

An acid–base indicator is made up of a weak acid (that has a particular colour) in equilibrium with its conjugate base (which has a different colour).

For example, consider methyl red indicator in Figure 1.16.

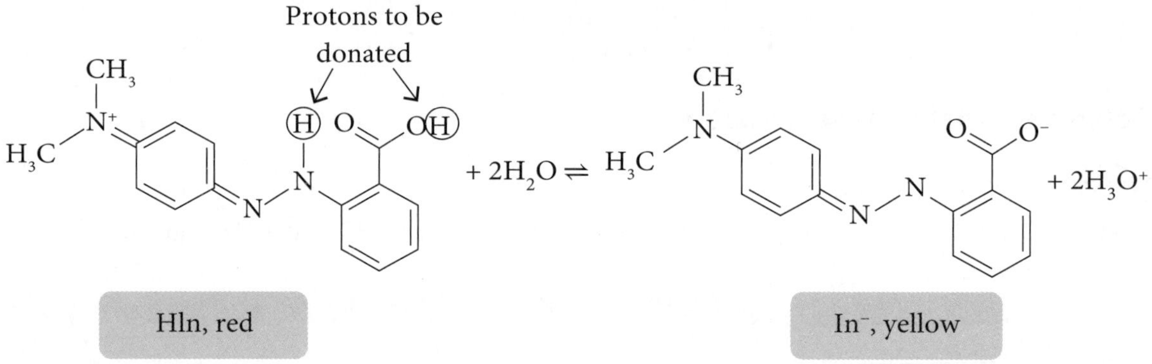

FIGURE 1.16 Methyl red indicator has a bright red colour (HIn) in acidic solution, which changes to the yellow conjugate base (In−).

These are complex chemical structures so, whilst it is not necessary to remember them, it *is* important to be able distinguish the acid form of the indicator from the base form. It is simply a matter of comparing the two structures and the number of H atoms each one has. In this example, the HIn, red form has more H atoms and is, therefore, the acid form.

When discussing their behaviour, it is easier to refer to the acid form of the indicator as HIn and the base form as In⁻.

These exist in an equilibrium system:

$$HIn(aq) + H_2O(l) \rightleftharpoons In^-(aq) + H_3O^+(aq)$$

Le Chatelier's principle can be used to explain how the indicator works in acidic and basic solutions.

In acidic solution

The system is surrounded by a very large excess of H_3O^+ ions.

According to Le Chatelier's principle, the system shifts to the left to try to remove the H_3O^+ ions. This is the HIn (red) side.

In basic solution

The system is surrounded by a very large excess of OH^- ions. These OH^- ions neutralise the H_3O^+ ions.

According to Le Chatelier's principle, the system shifts to the right to try to replenish the lost H_3O^+ ions. This is the In⁻ (yellow) side.

1.8.2 pH range of acid–base indicators

K and p*K*

K_a and K_b values can be extremely small. It is very useful to express these values more conveniently using the formula:

$$pK_a = -\log_{10} K_a$$

$$pK_b = -\log_{10} K_b$$

So, an acid such as hydrofluoric acid has a K_a value of 7.6×10^{-4}

$$pK_a = - \log_{10} 7.6 \times 10^{-4}$$

$$= \mathbf{3.12}$$

A base such as hydrazine has a K_b value of 1.3×10^{-6}

$$pK_b = - \log_{10} 1.3 \times 10^{-6}$$

$$= \mathbf{5.9}$$

pK_a of indicators

Acid–base indicators are very useful in volumetric analysis as end-point indicators in titrations (discussed on page 26).

If the indicator has been chosen correctly, then it will change colour when the pH of the solution is the same as the indicator's pKa:

pK$_a$(indicator) = pH(solution)

For example, the indicator bromophenol blue has a pK_a value of 4.2. When a solution containing this indicator reaches a pH of 4.2, it changes colour from yellow to blue.

The colour changes associated with indicators do not occur instantly at the pH that is equal to their pK_a value. Generally, they change colour over a narrow pH range:

pH range = p$K_a \pm 1$

1.9 Volumetric analysis

1.9.1 Acid–base titrations

In acid–base titrations we use known quantities of an indicator and acid or base to determine the unknown concentration of an acid or base.

The apparatus used to carry out acid–base titrations is shown in Figure 1.17.

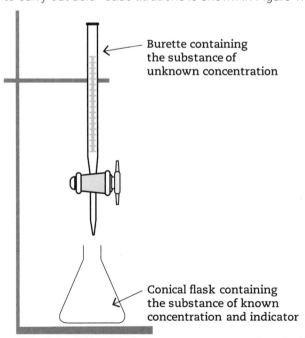

FIGURE 1.17 Apparatus used to carry out acid–base titrations

To determine the concentration of the base, first place it in a burette. The acid is placed in the conical flask along with a few drops of the indicator. The base is added from the burette until the end point, which is when the indicator changes colour.

> **Hint**
>
> **End point and equivalence point**
>
> It is important to distinguish between these two terms:
>
> The **equivalence point** of a titration is when JUST ENOUGH base has been added to neutralise the acid.
>
> The **end point** of a titration occurs when the pH of the solution causes the indicator present to change colour.

1.9.2 Titration curves

Titrations monitored using indicators have some limitations, for example, if one of the solutions has a pronounced colour.

Titrations that are monitored using pH meters can be more reliable. Monitoring is achieved by plotting a graph of pH against volume of base (or acid) added. This graph is called a **titration curve**.

The shape of a titration curve depends upon the nature of the acid and base involved (see Figures 1.18 to 1.21).

Strong base–strong acid

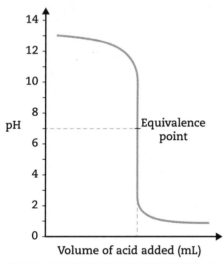

FIGURE 1.18 A titration curve for a strong base–strong acid titration

Strong base–weak acid

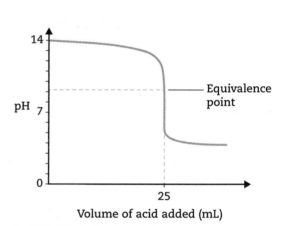

FIGURE 1.19 A titration curve for a strong base–weak acid titration

Weak base–strong acid

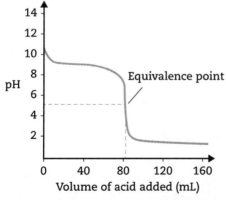

FIGURE 1.20 A titration curve for a weak base–strong acid titration

> **Hint**
>
> As weak base–weak acid titration curves have no clear equivalence point, they are not included in your exam.

Weak base–weak acid

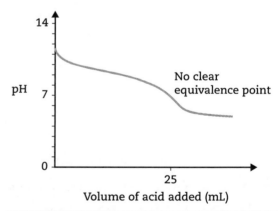

FIGURE 1.21 A titration curve for a weak base–weak acid titration

A lot can be learned about a titration by analysing its titration curve (see Figure 1.22).

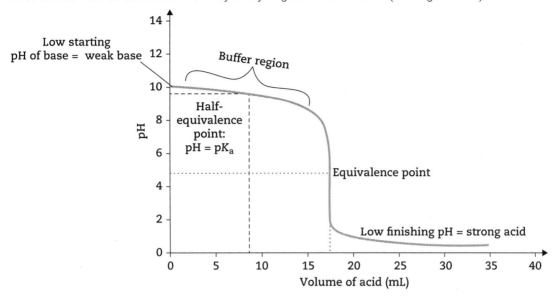

FIGURE 1.22 Features of a titration curve

Shape

Strong acids are obvious from their low pH, bases from their high pH.

Strong acids and bases tend to occupy the more horizontal parts of the curve.

Weak acids and bases give rise to the more diagonal parts.

Equivalence point

This is found by measuring half-way up the vertical part of the curve and reading across to the corresponding pH value (in the titration curve above, pH = 5).

Half-equivalence point

This is found by noting the corresponding volume at the equivalence point (17.5 mL), halving that value (8.75 mL) and drawing a line up to the curve at that volume. At the curve, a line is drawn across to the pH value (9.5).

This pH value at the half-equivalence point is *very* important. At this point, the pH = pK_a of the base. The pK_b of the base can be found using the equation:

$$pK_a + pK_b = 14$$

$$pK_b = 14 - pK_a$$

Buffer region

This is the section of the curve during which the pH of the solution changes little as a base is added. It is generally the 'slope' part of the graph.

In the graph above, a weak base such as ethylamine, $CH_3CH_2NH_2$, is being added to a strong acid such as HCl:

$$CH_3CH_2NH_2(aq) + HCl(aq) \rightleftharpoons CH_3CH_2NH_3^+(aq) + Cl^-(aq)$$

In this situation, where not much HCl has been added, a situation exists in which a weak base is in equilibrium with its conjugate acid:

$$CH_3CH_2NH_2(aq) \rightleftharpoons CH_3CH_2NH_3^+(aq)$$

This is a basic buffer solution.

Worked example

Question: Explain the shape of the curve between points X and Y on the titration curve below for the titration between propanoic acid, $CH_3CH_2COOH(aq)$, and sodium hydroxide, $NaOH(aq)$.

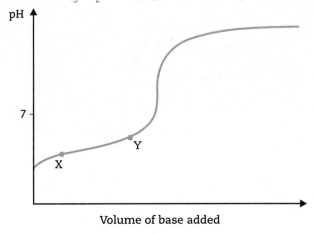

Step 1

Write a balanced equation.

$$CH_3CH_2COOH(aq) + NaOH \rightleftharpoons CH_3CH_2COO^-(aq) + Na^+(aq) + H_2O(l)$$

Step 2

Identify the region between points X and Y as the buffer region.

Step 3

Refer to balanced equation to identify the weak acid and its conjugate base.

$$\underset{\text{weak acid}}{CH_3CH_2COOH(aq)} \rightleftharpoons \underset{\text{conjugate base}}{CH_3CH_2COO^-(aq)}$$

Step 4

Identify the type of buffer solution (acid or base).

The area between points X and Y lies below pH 7 on the graph.

This is an acidic buffer region.

Explaining equivalence points

For a strong acid–strong base titration the equivalence point has a pH of 7, which is that of a neutral solution.

Why are the equivalence points of a weak acid–strong base and a strong acid–weak base *not* at pH 7?

Weak acid–strong base

Figure 1.19 (page 26) represents the titration curve. The equivalence point is at about pH 10.

This could be an ethanoic acid–sodium hydroxide titration:

$$CH_3COOH(aq) + NaOH(aq) \rightleftharpoons CH_3COO^-(aq) + Na^+(aq) + H_2O(l)$$

Looking at what remains in the conical flask at equivalence point:

No reactants

$Na^+(aq)$ = no acidic or basic properties

$H_2O(l)$ = no acidic or basic properties

$CH_3COO^-(aq)$ = the conjugate base of a weak acid = weak base, therefore pH = approximately 10.

Strong acid–weak base

Figure 1.20 (page 26) represents the titration curve. The equivalence point is at about pH 5.

This could be a hydrochloric acid–ammonia titration.

$$HCl(aq) + NH_3(aq) \rightleftharpoons NH_4^+(aq) + Cl^-(aq)$$

Looking at what remains in the conical flask at equivalence point:

No reactants

Cl^- = the conjugate base of a strong acid = no basic properties

$NH_4^+(aq)$ = the conjugate acid of a weak base, therefore pH is approximately 5.

1.9.3 Mathematical representations

Titration questions can be daunting but, with practice, they can become routine.

Worked example

Question: A standard solution of sodium carbonate, Na_2CO_3, was prepared by dissolving 1.318 g of sodium carbonate in 50 mL of water.

This solution was transferred to a 250.00 mL volumetric flask. Distilled water was added to the flask up to the 250.00 mL mark.

This solution was transferred to a burette.

A 20.00 mL sample of HCl of unknown concentration was pipetted into a conical flask.

A few drops of indicator were added, and the solution titrated with the sodium carbonate according to the equation below until the end point was reached. The volume of sodium carbonate added at the end point was 18.35 mL.

$$2HCl(aq) + Na_2CO_3(aq) \rightarrow 2NaCl(aq) + CO_2(g) + H_2O(l)$$

a Determine the concentration of the HCl solution.

b What indicator would have been used for this titration?

Part a

Step 1

Draw a diagram of the situation.

> **Hint**
> A diagram of the situation can help to keep track of the many steps in the calculation.

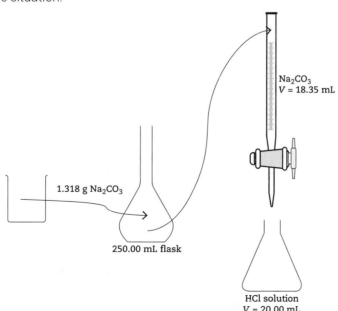

Step 2

Calculate the number of moles of Na_2CO_3 in 1.318 g.

> **Hint**
> **Key formula**
>
> $$\text{Number of moles, } n = \frac{m \text{ (given mass)}}{M \text{ (formula mass)}}$$

$$n = \frac{1.318}{106} = 0.0124 \text{ mol}$$

Step 3

Calculate the concentration of Na_2CO_3.

> **Hint**
> **Key formula**
>
> $$\text{Concentration, } c = \frac{n \text{ (number of moles)}}{V \text{ (volume of solution)}}$$

$$c = \frac{0.0124}{\frac{250.00}{1000}} = 0.0496 \text{ M}$$

> **Hint**
> When volumes are given in mL, the volume of the solution must *always* be divided by 1000.

Step 4

Calculate the number of moles of Na_2CO_3 added from the burette.

$$c = \frac{n}{V} \quad \therefore \quad n = c \times V = 0.0496 \times \frac{18.35}{1000} = 9.1 \times 10^{-4} \text{ mol}$$

Step 5

Refer to the balanced equation to determine the mole ratio between Na_2CO_3 and HCl.

2 moles HCl = 1 mole Na_2CO_3

For every mole of Na_2CO_3 added from the burette, there must have been 2 moles of HCl neutralised in the conical flask.

Step 6

Calculate the number of moles of HCl in the conical flask.

$$n(HCl) = 2 \times n(Na_2CO_3)$$

$$= 2 \times 9.1 \times 10^{-4}$$

$$= 1.82 \times 10^{-3} \text{ mol}$$

Step 7

Calculate the concentration of HCl.

$$n = 1.82 \times 10^{-3} \text{ mol}$$

$$V = 20.00 \text{ mL}$$

$$c = \frac{n}{V} = \frac{1.82 \times 10^{-8}}{\frac{20.00}{1000}} = 0.0905 \text{ M}$$

Part b

Step 1

Sketch the appropriate titration curve for this reaction – the reaction is a strong acid with a weak base.

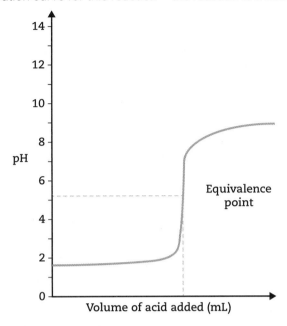

The pH of the equivalence point is about 5.

Hint

Key formula

Indicator range = pK_a (indicator) ± 1

Therefore, by referring to the table of acid–base indicators on page 14 of the *Chemistry formula and data booklet* available to you in your exam, the indicator should be:

Methyl red

This appears to be the most suitable indicator because it has a pK_a = 5.1, which is closest to the pH of the equivalence point.

Back titrations

These are an important type of titration because they allow analysis of substances that either do not react directly, are insoluble or are too volatile.

Worked example

Question: Chalk is a sedimentary rock made from seashells composed largely of calcium carbonate. A 10.00 g sample of chalk was dissolved in 100.00 mL of a 1.137 M solution of HCl. A 20.00 mL sample of this solution was transferred to a conical flask and titrated with a 0.184 M solution of NaOH. End point occurred at 21.35 mL. Determine the percentage of Ca^{2+} ions in the chalk sample.

The reaction between calcium carbonate and hydrochloric acid can be represented by the equation:

$$2HCl(aq) + CaCO_3(s) \rightarrow CaCl_2(aq) + CO_2(g) + H_2O(l)$$

Step 1

Draw a diagram similar to the set-up above.

Step 2

Calculate the initial number of moles of HCl.

$$n = 1.137 \times \frac{100.0}{1000} = 0.1137 \text{ mol}$$

Step 3

Calculate the number of moles of NaOH added from the burette.

$$n = 0.184 \times \frac{21.35}{1000} = 3.93 \times 10^{-3} \text{ mol}$$

This is equal to the number of moles of HCl in the 20.00 mL sample.

Step 4

Calculate the number of moles of HCl remaining after reaction with $CaCO_3$.
If 20.00 mL, sample contains 3.93×10^{-3} mol HCl;
therefore, a 100.00 mL sample contains $3.93 \times 10^{-3} \times 5 = 0.02$ mol HCl.

Step 5

Calculate the number of moles of HCl that reacted with the calcium carbonate.

$$n(\text{HCl that reacted}) = n(\text{initial HCl}) - n(\text{final HCl})$$
$$= 0.1137 - 0.02 = 0.0937 \text{ mol}$$

Step 6

Calculate the mass of $CaCO_3$.

$$m = n \times M$$

$$0.0937 \times 100 = 9.37 \text{ g}$$

Step 7

Calculate % of $CaCO_3 = Ca^{2+}$.

$$\% \text{ Ca}^{2+} = \frac{9.37}{10.00} \times 100 = 93.7\%$$

Glossary

amphiprotic
a substance that can gain or lose a hydrogen ion to act as an acid or a base

chemical change
a change that creates new substances with different physical and chemical properties to the reactants

closed system
no substances can enter or leave the system, but energy can be exchanged

conjugate acid
the acid that is formed when a base accepts a proton from an acid

conjugate base
the base that is formed when an acid donates a proton to a base

diprotic
a substance that can donate two protons to a base

dynamic equilibrium
a system in balance but where forwards and backwards reactions occur at the same rate

end point
the physical sign that indicates the equivalence point has been reached; in titration it is when the pH of the solution causes the indicator to change colour

equilibrium
a system in balance

equilibrium expression
the ratio of the concentration of products to reactants used to calculate the equilibrium constant

equivalence point
the point when the reactants are present in the ratio shown by the mole ratio in the balanced chemical equation for the reaction

heterogeneous system
a system where there is a mixture of phases of reactants and products

homogeneous system
a system where all the reactants and products are in the same phase

A+ DIGITAL FLASHCARDS
Revise this topic's key terms and concepts by scanning the QR code or typing the URL into your browser.

https://get.ga/aplus-qce-chem-u34

monoprotic
a substance that can donate one proton to a base; an acid that dissociates producing one hydrogen ion per molecule

open system
a system for which matter and energy can be exchanged with the surroundings

physical change
involves no new products

reaction quotient
the value of the equilibrium expression when calculated, Q

reversible
describes a reaction where the products can be converted back to the reactants

self-ionisation constant (K_w)
the equilibrium expression that shows water ionising into two ions

stoichiometry
the process of using the relationships of substances in a chemical reaction to determine quantitative data. Literally means the 'measure of elements'

thermochemical equation
a balanced equation with its corresponding enthalpy changed

titration curve
a graphical representation of a titration with the x-axis showing the volume of titrant added and the y-axis showing the pH of the solution

triprotic
a substance that can donate three protons to a base

9780170459150

Revision summary

Use the following summary of syllabus dot points and key knowledge within Unit 3 Topic 1 to ensure that you have thoroughly reviewed the content. Provide a brief definition or comment for each item to demonstrate your understanding or code them using the traffic light system – Green (all good); Amber (needs some review); Red (priority area to review).

Chemical equilibrium	
• recognise that chemical systems may be open (allowing matter and energy to be exchanged with the surroundings) or closed (allow energy, but not matter, to be exchanged with the surroundings)	
• understand that physical changes are usually reversible, whereas only some chemical reactions are reversible	
• appreciate that observable changes in chemical reactions and physical changes can be described and explained at an atomic and molecular level	
• symbolise equilibrium equations by using ⇌ in balanced chemical equations	
• understand that, over time, physical changes and reversible chemical reactions reach a state of dynamic equilibrium in a closed system, with the relative concentrations of products and reactants defining the position of equilibrium	
• explain the reversibility of chemical reactions by considering the activation energies of the forward and reverse reactions	
• analyse experimental data, including constructing and using appropriate graphical representations of relative changes in the concentration of reactants and products against time, to identify the position of equilibrium.	

⟩⟩

Factors that affect equilibrium	
• explain and predict the effect of temperature change on chemical systems at equilibrium by considering the enthalpy change for the forward and reverse reactions	
• explain the effect of changes of concentration and pressure on chemical systems at equilibrium by applying collision theory to the forward and reverse reactions	
• apply Le Chatelier's principle to predict the effect changes of temperature, concentration of chemicals, pressure and the addition of a catalyst have on the position of equilibrium and on the value of the equilibrium constant.	

Equilibrium constants	
• understand that equilibrium law expressions can be written for homogeneous and heterogeneous systems and that the equilibrium constant (K_c), at any given temperature, indicates the relationship between product and reactant concentrations at equilibrium	
• deduce the equilibrium law expression from the equation for a homogeneous reaction and use equilibrium constants (K_c) to predict qualitatively the relative amounts of reactants and products (equilibrium position)	
• deduce the extent of a reaction from the magnitude of the equilibrium constant	
• use appropriate mathematical representation to solve problems, including calculating equilibrium constants and the concentration of reactants and products.	

››	Properties of acids and bases	
	• understand that acids are substances that can act as proton (hydrogen ion) donors and can be classified as monoprotic or polyprotic depending on the number of protons donated by each molecule of the acid	
	• distinguish between strong and weak acids and bases in terms of the extent of dissociation, reaction with water and electrical conductivity and distinguish between the terms strong and concentrated for acids and bases.	
	pH scale	
	• understand that water is a weak electrolyte and the self-ionisation of water is represented by $K_w = [H^+][OH^-]$; K_w can be used to calculate the concentration of hydrogen ions from the concentration of hydroxide ions in a solution	
	• understand that the pH scale is a logarithmic scale and the pH of a solution can be calculated from the concentration of hydrogen ions using the relationship $pH = -\log_{10}[H^+]$	
	• use appropriate mathematical representation to solve problems for hydrogen ion concentration $[H^+(aq)]$, pH, hydroxide ion concentrations $[OH^-(aq)]$ and pOH.	
	Brønsted–Lowry model	
	• recognise that the relationship between acids and bases in equilibrium systems can be explained using the Brønsted–Lowry model and represented using chemical equations that illustrate the transfer of hydrogen ions (protons) between conjugate acid–base pairs	››

9780170459150

»

• recognise that amphiprotic species can act as Brønsted–Lowry acids and bases	
• identify and deduce the formula of the conjugate acid (or base) of any Brønsted–Lowry base (or acid)	
• appreciate that buffers are solutions that are conjugate in nature and resist a change in pH when a small amount of an acid or base is added; Le Chatelier's principle can be applied to predict how buffer solutions respond to the addition of hydrogen ions and hydroxide ions.	

Dissociation constants

• recognise that the strength of acids is explained by the degree of ionisation at equilibrium in aqueous solution, which can be represented with chemical equations and equilibrium constants (K_a)	
• determine the expression for the dissociation constant for weak acids (K_a) and weak bases (K_b) from balanced chemical equations	
• analyse experimental data to determine and compare the relative strengths of acids and bases	
• use appropriate mathematical representation to solve problems, including calculating dissociation constants (K_a and K_b) and the concentration of reactants and products.	

Acid–base indicators

• understand that an acid–base indicator is a weak acid or a weak base where the components of the conjugate acid–base pair have different colours; the acidic form is of a different colour to the basic form	

»

››	• explain the relationship between the pH range of an acid–base indicator and its pK_a value	
	• recognise that indicators change colour when the pH = pK_a and identify an appropriate indicator for a titration, given equivalence point of the titration and pH range of the indicator.	
	Volumetric analysis	
	• distinguish between the terms end point and equivalence point	
	• recognise that acid–base titrations rely on the identification of an equivalence point by measuring the associated change in pH, using chemical indicators or pH meters, to reveal an observable end point	
	• sketch the general shapes of graphs of pH against volume (titration curves) involving strong and weak acids and bases. Identify and explain their important features, including the intercept with pH axis, equivalence point, buffer region and points where pK_a = pH or pK_b = pOH	
	• use appropriate mathematical representations and analyse experimental data and titration curves to solve problems and make predictions, including using the mole concept to calculate moles, mass, volume and concentration from volumetric analysis data.	
	• Mandatory practical: Acid–base titration to calculate the concentration of a solution with reference to a standard solution.	

Exam practice

Topic 1: Chemical equilibrium systems

Multiple-choice questions

Each multiple-choice question is worth 1 mark.

Solutions start on page 167.

Question 1

The industrial production of ammonia, NH_3, is given by the equation:

$$N_2(g) + 3H_2(g) \rightleftharpoons 2NH_3(g) \quad \Delta H = -92 \, kJ \, mol^{-1}$$

The graph below represents the energy profile for this equilibrium system.

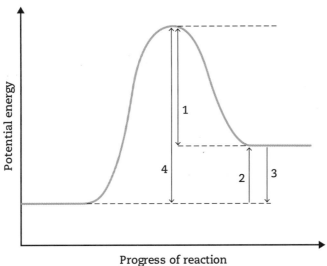

Identify the number on the graph that represents the heat of reaction for the equation as written above.

A 1

B 2

C 3

D 4

Question 2

The production of $PCl_5(g)$ can be represented by the equation:

$$PCl_3(g) + Cl_2(g) \rightleftharpoons PCl_5(g) \quad \Delta H = -98 \, kJ \, mol^{-1}$$

The yield of PCl_5 can be increased by

A increasing the temperature of the system.

B removing $Cl_2(g)$ from the system.

C adding a catalyst.

D increasing the pressure of the system.

Question 3

The production of hydrogen iodide, HI, is given by the equation:

$$H_2(g) + I_2(g) \rightleftharpoons 2HI(g)$$

The K_c value for this reaction at 200°C is 8.43.

A mixture of $H_2(g)$ and $I_2(g)$ was placed in a container at 200°C. After a few minutes, the concentrations of the species in the container were measured and found to be

- $[H_2] = 0.25\,M$
- $[I_2] = 0.50\,M$
- $[HI] = 0.65\,M$.

Which one of the following statements is true about this system?

A The value of Q is greater than K_c and the system will shift right.

B The value of Q is less than K_c and the system will shift left.

C The value of Q is greater than K_c and the system will shift left.

D The value of Q is less than K_c and the system will shift right.

Question 4

Identify the acid in the list below that would have the highest electrical conductivity.

A Sulfuric acid

B Carbonic acid

C Hydrochloric acid

D Ethanoic acid

Question 5

Calculate the pH of a 0.36 M solution of sulfuric acid.

A 0.44

B 0.88

C 0.28

D 0.14

Question 6

If a small amount of NaOH is added to the buffer solution made up of ethanamine and ethylammonium chloride, then the system shifts to the

A left to reduce the number of ethylammonium ions.

B right to reduce the amount of ethanamine.

C left to increase the amount of ethanamine.

D right to increase the number of ethylammonium ions.

Question 7

If the K_a value for cyanic acid, HCNO, is 3.5×10^{-4}, calculate the pH of a 0.05 M solution.

A 1.19

B 1.72

C 2.38

D 4.76

Question 8

The reaction between nitric acid and diethylamine is represented by the equation:

$$HCl(aq) + (CH_3)_2NH(aq) \rightarrow (CH_3)_2NH_2^+(aq) + Cl^-(aq)$$

If a titration is carried out between these reactants, which indicator from the table below would be most suitable?

	Indicator	pK_a
A	Malachite green	1.3
B	Bromophenol blue	4.0
C	Thymol blue	8.8
D	Alizarin yellow	11.0

Question 9

Potassium hydrogen phthalate, KHP, and sodium hydroxide react according to the balanced equation:

$$KHC_8H_4O_4(aq) + NaOH(aq) \rightarrow KNaC_8H_4O_4(aq) + H_2O(l)$$

A sample of KHP is weighed and dissolved in 50 mL of deionised water. This is transferred to a 250.00 mL volumetric flask and made up to the mark with deionised water. A 20.00 mL sample of this solution was titrated with a 0.139 M solution of NaOH. The end point occurred at 23.45 mL. Calculate the mass of KHP that was weighed.

A 0.326 g

B 4.084 g

C 8.169 g

D 0.665 g

Question 10

Given that the pK_b value for aniline, $C_6H_5NH_2$, is 9.37, calculate the pH of a 0.826 M solution of aniline chloride.

A 0.3

B 1.2

C 2.4

D 4.6

Question 11

Propanoic acid dissociates according to the equation:

$$CH_3CH_2COOH(aq) + H_2O(l) \rightleftharpoons CH_3CH_2COO^-(aq) + H_3O^+(l)$$

Calculate the percentage ionisation of propanoic acid in a 0.185 M solution that has a pH of 2.8.

A 2.65%

B 0.86%

C 0.92%

D 0.30%

Question 12

Which of the following shows the correct dissociation products in the correct order for the triprotic acid boric acid, H_3BO_3?

A $HBO_3^{2-}, H_2BO_3^-, BO_3^{3-}$

B $BO_3^{3-}, HBO_3^{2-}, H_2BO_3^-$

C $H_2BO_3^-, HBO_3^{2-}, BO_3^{3-}$

D $H_3BO_3, H_2BO_3^-, HBO_3^{2-}$

Question 13

Phosphorus pentachloride, PCl_5, decomposes according to the equation:

$$PCl_5(g) \rightleftharpoons PCl_3(g) + Cl_2(g) \qquad \Delta H = +98\,kJ\,mol^{-1}$$

The graph below shows the changes in concentration of each species as changes were made to the system.

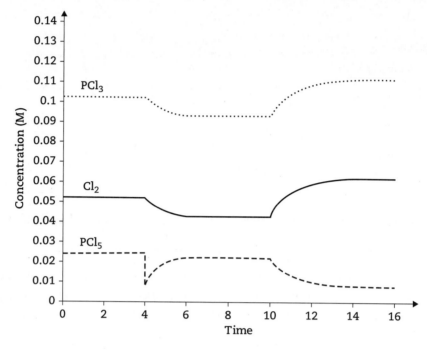

Identify the change made to the system at time $t = 10$ min.

A   Temperature was increased, forcing the system to shift right.

B   Pressure was increased, forcing the system to shift right.

C   Temperature was decreased, forcing the system to shift right.

D   Some $PCl_5(g)$ was removed from the system, forcing the system to shift right.

Question 14

A titration is carried out between hydrochloric acid and an unknown weak base. The progress of the reaction was recorded in the titration curve below.

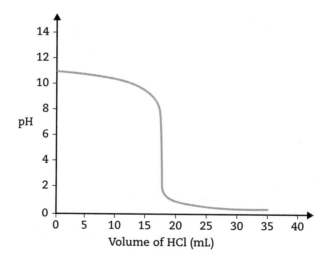

From the curve, determine the K_b of the base.

A 5.0×10^{-11}

B 8.0×10^{-4}

C 2.5×10^{-9}

D 2.0×10^{-4}

Short response questions

Solutions start on page 168.

Question 15 (4 marks)

The reaction between nitrogen gas and oxygen gas to produce nitric oxide proceeds according to the equation:

$$N_2(g) + O_2(g) \rightleftharpoons 2NO(g) \quad K_c = 3.4 \times 10^{-4} \text{ at } 440\,K$$

a If the initial concentrations of the reactants at 440 K were 0.15 M for N_2 and 0.15 M for O_2, calculate the equilibrium concentration of NO.   2 marks

If the temperature is increased to 540 K and the system allowed to come to equilibrium, the value of K_c is 1.6×10^{-7}.

b Deduce whether this is an endothermic or an exothermic reaction, giving a reason for your decision.   2 marks

Question 16 (5 marks)

Methyl orange is commonly used to indicate the end point in acid bases titrations. It exists as one of two structural forms depending on the pH of the solution it is in. It has a pK_a value of 4.1. Its structures are shown below.

a With reference to Le Chatelier's principle, and correctly identifying structures A and B as acidic or basic, describe how a methyl orange indicator changes colour from red in an acidic solution to yellow at the end point in an acid–base titration.   3 marks

b The general equation for the dissociation of an acid–base indicator is given by:

$$HIn(aq) + H_2O(l) \rightleftharpoons In^-(aq) + H_3O^+(aq)$$

Use this information to explain why, at the equivalence point in a titration

$$pK(\text{indicator}) = pH$$

2 marks

Question 17 (5 marks)

The reaction between pentanoic acid and sodium hydroxide is represented by the balanced equation:

$$CH_3(CH_2)_3COOH(aq) + NaOH(aq) \rightarrow CH_3(CH_2)_3COO^-(aq) + Na^+(aq) + H_2O(l)$$

$$K_a(\text{pentanoic acid}) = 1.44 \times 10^{-5}$$

A mixture is made up by adding 84.35 mL of a 1.85 M solution of pentanoic acid with 113.45 mL of a 0.92 M NaOH solution.

Calculate the pH of this mixture. Show all working.

Question 18 (7 marks)

The titration between hydrochloric acid and methanamine is represented by the equation:

$$HCl(aq) + CH_3NH_2(aq) \rightarrow CH_3NH_3^+(aq) + Cl^-(aq)$$

The progress of the titration can be monitored using a titration curve as shown in the following graph.

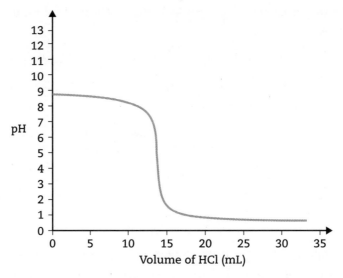

a Identify the buffer region and the half-equivalence point by labelling them on the graph above.   2 marks

b Determine the pK_b of methylamine.   1 mark

c Explain, with the aid of equations, why the buffer region is so named.   2 marks

d Explain why the pH of the equivalence point is *not* pH 7.   2 marks

Question 19 (5 marks)

The reaction between sulfur dioxide and oxygen to produce sulfur trioxide can be represented by the balanced equation:

$$2SO_2(g) + O_2(g) \rightleftharpoons 2SO_3(g) \quad \Delta H = -93\,kJ\,mol^{-1}$$

The progress of the reaction is shown on the concentration versus time graph below.

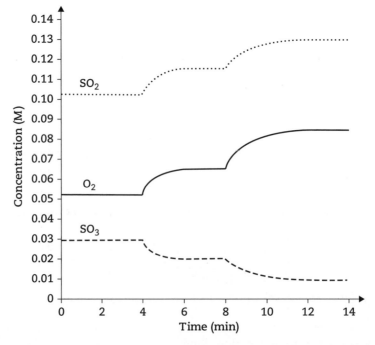

Analyse the data at times $t = 6$ min and $t = 12$ min, then write a statement relating the magnitude of the equilibrium constant and temperature. Justify your reasoning with appropriate calculations.

Question 20 (6 marks)

Nitrogen is an important component of many fertilisers. It is usually present in the form of ammonium chloride or nitrate. To determine the percentage of nitrogen in a fertiliser, a student dissolved a 0.504 g sample of the fertiliser in 30.00 mL of a 0.284 M solution of NaOH. The mixture was heated gently to drive off the ammonia formed according to the equation:

$$NH_4NO_3(aq) + NaOH(aq) \rightarrow NaNO_3(aq) + H_2O(l) + NH_3(g)$$

This mixture was titrated with a 0.196 M solution of HCl. The end point occurred at 17.60 mL. Determine the percentage, by mass, of the nitrogen in the fertiliser.

Chapter 2
Topic 2: Oxidation and reduction

Topic summary

Oxidation–reduction reactions make up one of the largest groups of chemical reactions, including many of the most industrially important reactions and reactions necessary for our existence.

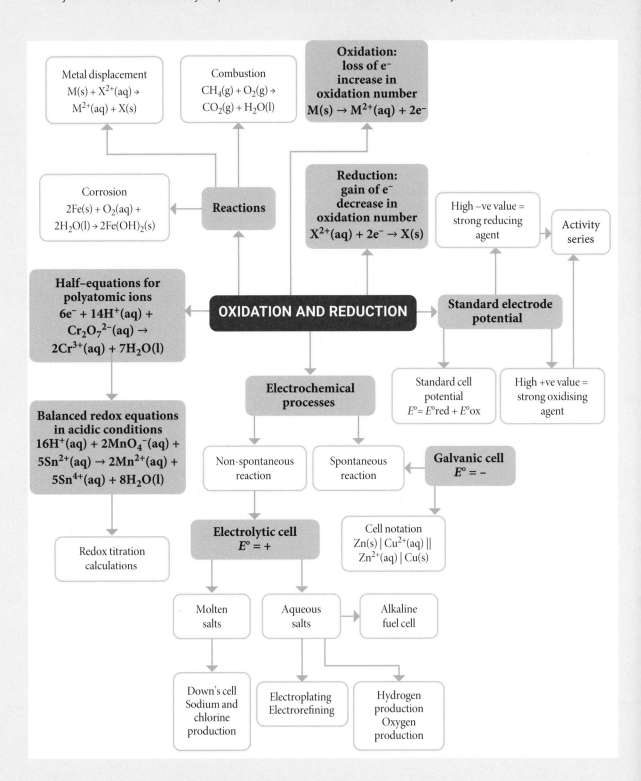

2.1 Redox reactions

2.1.1 The range of redox reactions

Oxidation–reduction reactions or **redox reactions** involve the transfer of electrons, and come in many forms.

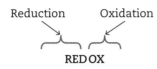

FIGURE 2.1 Redox

U3 – TOPIC 2

Metal displacement reactions

The most common forms of redox reactions occur when a piece of metal is added to an ionic solution containing the ion of another metal:

$$Zn(s) + CuSO_4(aq) \rightarrow ZnSO_4(aq) + Cu(s)$$

The SO_4^{2-} ion remains unchanged throughout the reaction. It is called a spectator ion and can be discarded for the purposes of this discussion.

Consider the metals in this reaction.

Zinc:
Zn(s) reacts to become Zn^{2+}(aq) ions:

$$Zn(s) \rightarrow Zn^{2+}(aq)$$

$$Zn(s) \rightarrow Zn^{2+}(aq) + 2e^-$$

The Zn has *lost* two electrons – it has been *oxidised*. This is the oxidation half-equation.

Copper:
Cu^{2+}(aq) ions react to become Cu(s):

$$Cu^{2+}(aq) + 2e^- \rightarrow Cu(s)$$

The Cu^{2+} has *gained* two electrons – it has been *reduced*. This is the reduction half-equation.

When the half-equations are combined, a net ionic equation is created:

$$Zn(s) + Cu^{2+}(aq) \rightarrow Zn^{2+}(aq) + Cu(s)$$

> **Hint**
>
> Rule #1
>
> Balance for charge by adding the appropriate number of e^- (electrons) to the more positive side of the equation.

Worked example

Question: If a piece of copper metal is placed in a silver nitrate solution, eventually the copper becomes coated with small crystals of silver metal.

a Write the oxidation and reduction half-equations for this reaction.

b Write the net ionic equation.

Step 1

Write a balanced equation for the reaction.

$$Cu(s) + AgNO_3(aq) \rightarrow Cu(NO_3)_2(aq) + Ag(s)$$

Step 2

Identify the oxidation and reduction half-equations.

Oxidation: $\quad Cu(s) \rightarrow Cu^{2+}(aq) + 2e^-$

Reduction: $\quad Ag^+(aq) + e^- \rightarrow Ag(s)$

Step 3

Combine the half-equations.

The number of electrons lost *must* equal those gained.

The number of e⁻ in the reduction half-equation must be multiplied by 2.

$$2Ag^+(aq) + 2e^- \rightarrow 2Ag(s)$$

Combining the half-equations and cancelling the electrons gives:

$$Cu(s) + 2Ag^+(aq) \rightarrow Cu^{2+}(aq) + 2Ag(s)$$

Combustion

Consider a piece of magnesium ribbon burning in oxygen:

$$2Mg(s) + O_2(g) \rightarrow 2MgO(s)$$

Splitting this into the half-equations, we see that the Mg is oxidised, while oxygen is reduced:

$$2Mg(s) \rightarrow 2Mg^{2+}(s) + 4e^-$$

$$O_2(g) + 4e^- \rightarrow 2O^{2-}(s)$$

Corrosion

In its simplest form, corrosion is the oxidation of a metal in air to produce the metal oxide:

$$4Fe(s) + 3O_2(g) \rightarrow 2Fe_2O_3(s)$$

This is complicated by the presence of water, which acts as an electrolyte. Electrolytes are able to conduct the flow of electrons.

The corrosion half-equations for iron become:

Oxidation: $$Fe(s) \rightarrow Fe^{2+}(aq) + 2e^-$$

Reduction: $$O_2(g) + 2H_2O(l) \rightarrow 4OH^-(aq)$$

$$2Fe(s) + O_2(g) + 2H_2O(l) \rightarrow 2Fe(OH)_2(s)$$

The $Fe(OH)_2$ precipitate oxidises further to produce hydrated iron(III) hydroxide – rust:

$$4Fe(OH)_2(s) + O_2(g) \rightarrow 2(Fe_2O_3.H_2O)(s) + 2H_2O(l)$$

Electrochemical processes

An electrochemical process is a chemical reaction that causes or is caused by the flow of electrons.

Remember the metal displacement reaction:

$$Zn(s) + Cu^{2+}(aq) \rightarrow Zn^{2+}(aq) + Cu(s)$$

If the two half-reactions can be chemically separated but still connected by a wire, the energy of the electrons that move from the zinc to the copper can be harnessed:

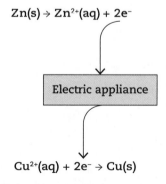

FIGURE 2.2 Harnessing electricity from a spontaneous reaction

This is the basis of galvanic cells; that is, batteries.

If a spontaneous reaction like the one just discussed can produce electricity, can the opposite occur – a non-spontaneous reaction being made to work by a supplied electric current?

This is the basis of the electrolytic cell.

> **Hint**
>
> A spontaneous reaction is one that proceeds left to right, as written, from reactants to products.
>
> A non-spontaneous reaction is one that does not proceed from left to right.

Consider the reaction:

$$2NaCl(l) \rightarrow 2Na^+(l) + Cl_2(g)$$

$$Na^+(s) + e^- \rightarrow Na(l)$$

$$2Cl^-(s) \rightarrow Cl_2(g) + 2e^-$$

This would never occur spontaneously because the Na^+ and Cl^- ions are far too stable.

2.1.2 Predicting loss and gain of electrons

The periodic table can be used to assist with predicting which elements are oxidised and which are reduced.

Fundamentally, the periodic table can be split into two: metals and non-metals. Metals lose electrons; they are oxidised. Non-metals gain electrons; they are reduced.

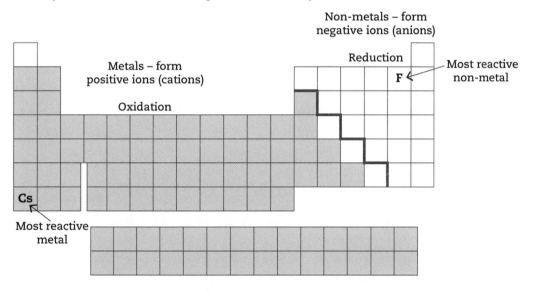

FIGURE 2.3 Blank periodic table, showing the location of metals and non-metals

The reactivity of an element is a measure of how easily the element loses electrons (oxidation) or gains electrons (reduction).

When discussing the reactivity of elements, three factors need to be considered:

- number of protons (p^+) in the nucleus
- distance of valence electrons from the nucleus
- number of shells between valence electrons and the nucleus.

Ionisation energy

Ionisation energy is the amount of energy required to remove an electron from a gaseous species:

$$E(g) \rightarrow E^+(g) + e^-$$

This is the *first* ionisation energy.

The *second* ionisation energy is given by:

$$E^+(g) \rightarrow E^{2+}(g) + e^-$$

Worked example

Question: Explain why the first ionisation energy of caesium is 375.71 kJ mol⁻¹, whereas for sodium it is 486 kJ mol⁻¹.

Even though there are many more protons in the nucleus of Cs than Na, the outer (valence) electron of Cs is much further away and there are many more electron shells between the nucleus and the valence electron (electron shielding).

Therefore, the valence electron of Cs is held only very weakly by the distant nucleus. It requires very little energy to remove the valence electron from Cs.

Therefore, its first ionisation energy is smaller than for Na.

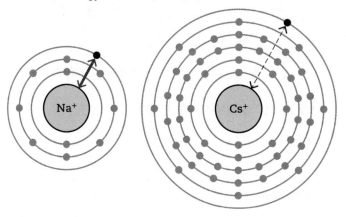

FIGURE 2.4 Comparison of the strength of the attraction of the nucleus for the valence electron for sodium and caesium

Ionisation energy trends in the periodic table

> **Hint**
> **Key concept**
>
> Elements react by losing or gaining electrons to attain a full outer shell of electrons.

As the periodic table moves from left to right, more electrons are added to the valence shell.

The first element in period 2 (the second row) has one valence electron. Beryllium has two and so on.

When five electrons are added, in the case of nitrogen, it becomes easier to gain three electrons rather than to lose the five valence electrons.

Worked example

Question: Explain why fluorine is much more reactive than iodine, even though its first ionisation energy is much higher (1681 kJ mol⁻¹ compared with 1008 kJ mol⁻¹ for iodine).

Even though fluorine has fewer protons in its nucleus, there are far fewer electron shells between it and nearby electrons. It will exert a much stronger attraction for nearby electrons, making it much more reactive than iodine. In iodine, the more positive nucleus is shielded by many electron shells, making its attractive power much weaker.

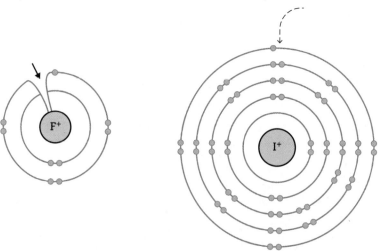

FIGURE 2.5 Comparison of the strength of the attraction of the nucleus and electrons from other species for fluorine and iodine

The situations here regarding caesium and fluorine are extreme examples of reactivity in the periodic table. Many factors need to be considered when discussing the relative reactivities of metals and non-metals and are beyond the scope of this course.

2.1.3 Identification in redox reactions

A number of definitions exist to describe oxidation and reduction, but in their simplest terms:

Oxidation is the **lO**ss of electrons
Reduction is, therefore, the gain of electrons.

Redox terminology

Consider the redox reaction:

$$Zn(s) + Cu^{2+}(aq) \rightarrow Zn^{2+}(aq) + Cu(s)$$

In this reaction, Zn has lost electrons – it has been oxidised.
Cu^{2+} has gained electrons – it has been reduced.
Cu^{2+} is the element that caused $Zn(s)$ to oxidise; therefore, it is the **oxidising agent** or **oxidant**.
Zn is the element that caused Cu^{2+} to be reduced; therefore, it is the **reducing agent** or **reductant**.

2.1.4 Oxidation

By keeping track of the electrons lost or gained, it is possible to deal with more complex situations, as in the worked example below.

Worked example

Question: Write the oxidation and reduction half-equations and the net ionic equation for the reaction between magnesium metal and nitrogen gas to give magnesium nitride powder.

$$4Mg(s) + 3N_2(g) \rightarrow 2Mg_2N_3(s)$$

Step 1

Track the progress of Mg.

$$Mg(s) \rightarrow Mg^{2+}(s)$$

Step 2

Use Rule #1 to balance for charge by adding e^- to the more positive side.

Hint

Rule #1

Balance for charge by adding the appropriate number of e^- (electrons) to the more positive side of the equation.

$$Mg(s) \rightarrow Mg^{2+}(s) + \mathbf{2e^-}$$

Step 3

Track the progress of N_2.

$$N_2(g) \rightarrow 2N^{3-}(s)$$

Step 4

Use Rule #1.

$$N_2(g) + \mathbf{6e^-} \rightarrow 2N^{3-}(s)$$

Step 5

The number of e^- in each equation must be the same, so the number of electrons in the Mg half-equation must be multiplied by 3 and Rule #2 used.

> **Hint**
>
> **Rule #2**
>
> If the number of e^- in a half-equation is multiplied by a number, then EVERYTHING in the half-equation must be multiplied by that number.

$$3Mg(s) \rightarrow 3Mg^{2+}(s) + 6e^-$$

$$N_2(g) + 6e^- \rightarrow 2N^{3-}(s)$$

Step 6

Cancelling the $6e^-$ in both half-equations and adding them together gives:

$$3Mg(s) + N_2(g) \rightarrow 3Mg^{2+}(s) + 2N^{3-}(s)$$

This can be verified by writing the overall neutral species equation:

$$3Mg(s) + N_2(g) \rightarrow Mg_3N_2(s)$$

Oxidation half-equations for polyatomic ions

Worked example

Question: Write the half-equation for the reduction of dichromate ions, $Cr_2O_7^{2-}$(aq), to chromium ions, Cr^{3+}(aq), in an acidic solution.

$$Cr_2O_7^{2-}(aq) \rightarrow Cr^{3+}(aq)$$

$$Cr_2O_7^{2-}(aq) \rightarrow 2Cr^{3+}(aq)$$

$$Cr_2O_7^{2-}(aq) \rightarrow 2Cr^{3+}(aq) + 7H_2O(l)$$

$$14H^+(aq) + Cr_2O_7^{2-}(aq) \rightarrow 2Cr^{3+}(aq) + 7H_2O(l)$$

$$\underbrace{14H^+(aq) + Cr_2O_7^{2-}(aq)}_{\text{overall 12+ charge}} \rightarrow \underbrace{2Cr^{3+}(aq)}_{\text{overall 6+ charge}} + 7H_2O(l)$$

$$6e^- + 14H^+(aq) + Cr_2O_7^{2-}(aq) \rightarrow 2Cr^{3+}(aq) + 7H_2O(l)$$

The half-equation for the reduction of dichromate ions is:

$$6e^- + 14H^+(aq) + Cr_2O_7^{2-}(aq) \rightarrow 2Cr^{3+}(aq) + 7H_2O(l)$$

H^+ ions form part of the reactants in this system. These are always referred to as taking place in acidic solution.

Similar reactions take place in basic (OH^-) solution, but they are beyond the scope of this course.

Redox reactions in acidic solution

The rules employed in the example above can be used when determining equations for redox reactions in acidic solution.

Worked example

Question: Write a balanced equation for the oxidation in acidic solution of Sn^{2+} to Sn^{4+} by MnO_4^- ions, which are reduced to Mn^{2+} ions.

Step 1

> **Hint**
> Rule #3
> Write the partial or skeleton half-equation.

$$MnO_4^-(aq) \rightarrow Mn^{2+}(aq)$$

$$Sn^{2+}(aq) \rightarrow Sn^{4+}(aq)$$

Step 2

> **Hint**
> Rule #4
> Balance for elements other than H and O.

(no change)

Step 3

> **Hint**
> Rule #5
> Balance for O by adding the appropriate number of H_2O molecules to the opposite side.

$$MnO_4^-(aq) \rightarrow Mn^{2+}(aq) + \mathbf{4H_2O(l)}$$

$$Sn^{2+}(aq) \rightarrow Sn^{4+}(aq)$$

Step 4

> **Hint**
> Rule #6
> Balance for H by adding the appropriate number of H^+ ions to the opposite side.

$$\mathbf{8H^+(aq)} + MnO_4^-(aq) \rightarrow Mn^{2+}(aq) + 4H_2O(l)$$

$$Sn^{2+}(aq) \rightarrow Sn^{4+}(aq)$$

Step 5

> **Hint**
> Rule #1
> Balance for charge by adding the appropriate number of e^- (electrons) to the more positive side of the equation.

$$\mathbf{5e^-} + 8H^+(aq) + MnO_4^-(aq) \rightarrow Mn^{2+}(aq) + 4H_2O(l)$$

$$Sn^{2+}(aq) \rightarrow Sn^{4+}(aq) + \mathbf{2e^-}$$

Step 6

> **Hint**
> Rule #2
> If the number of e^- in a half-equation is multiplied by a number, then EVERYTHING in the half-equation must be multiplied by that number.

$$\mathbf{10e^-} + \mathbf{16}H^+(aq) + \mathbf{2}MnO_4^-(aq) \rightarrow \mathbf{2}Mn^{2+}(aq) + \mathbf{8H_2O(l)} \qquad (\times 2)$$

$$\mathbf{5}Sn^{2+}(aq) \rightarrow \mathbf{5}Sn^{4+}(aq) + \mathbf{10e^-} \qquad (\times 5)$$

Cancelling the $10e^-$ in both half-equations and adding them together gives:

$$16H^+(aq) + 2MnO_4^-(aq) + 5Sn^{2+}(aq) \rightarrow 2Mn^{2+}(aq) + 5Sn^{4+}(aq) + 8H_2O(l)$$

> **Hint**
> In an exam, it can be useful to circle or underline the final equation to assist an exam marker.

Deducing the oxidation state

The **oxidation state** of an elemental ion is straightforward – it is the charge on the ion.

So, the oxidation state of the Mg^{2+} ion is +2.

The oxidation state of the Br^- ion is −1.

Determining the oxidation state of an element in a polyatomic ion is more challenging.

The following rules can be used when determining oxidation states.

1   The oxidation state of any element is zero.

2   The oxidation state of a monatomic ion is the same as the charge on that ion.

3   The oxidation state of hydrogen in compounds is always +1, except when forming metal hydrides, when it is −1.

4   The oxidation state of oxygen in any compound is always −2, except when forming peroxides, where it is −1, and in F_2O, where it is +2.

5   In a neutral compound, the sum of the oxidation states must always equal zero.

6   In polyatomic ions, the sum of the oxidation states must always equal the charge on the ion.

7   The most electronegative species in a compound is assigned the negative oxidation state; the less electronegative species is assigned the positive oxidation state.

Worked example

Question: Determine the oxidation state of chromium in the dichromate ion, $Cr_2O_7^{2-}$.

Step 1 (using rule #4)

O = −2

Therefore, charge due to the seven oxygen atoms = −2 × 7 = −14.

Step 2 (using rule #6)

$$-2 = (2 \times Cr) + (-14)$$
$$Cr = +6$$

Oxidation states of transition metals

Main group metals (those in groups 1, 2 and 13, such as sodium, magnesium and aluminium respectively), have predictable oxidation states.

However, transition metals can have varying oxidation states. Iron can have +2 or +3 oxidation states. Chromium can have +2, +3 or +6. Therefore, when referring to a transition metal compound, it is important to refer to the oxidation state of the metal. This is done using roman numerals. $FeCl_2$ is iron(II) chloride, $FeCl_3$ is iron(III) chloride. Iron(III) oxide has the formula F_2O_3.

Identifying redox reactions using oxidation states

So far, oxidation and reduction have been defined by loss and gain of electrons. Sometimes, particularly in complex reactions, identifying oxidised and reduced species can be difficult.

A far more useful characterisation of oxidation and reduction is by inspecting how oxidation states change.

OXIDATION is the **INCREASE** in oxidation state.

REDUCTION is the **DECREASE** in oxidation state.

Worked example

Question: Determine if the reaction below is a redox reaction.

$$H_2SO_4(aq) + 2NaOH(aq) \rightarrow Na_2SO_4(aq) + 2H_2O(l)$$

Oxidation state of Na in NaOH = +1

Oxidation state of Na in Na_2SO_4 = +1

No change in oxidation state of Na.

This is *not* a redox reaction.

Worked example

Question: Determine if the reaction below is a redox reaction.

$$Cl^-(aq) + 3SO_4^{2-}(aq) + 6H^+(aq) \rightarrow ClO_3^-(aq) + 3SO_2(g) + 3H_2O(l)$$

Oxidation state of Cl in Cl^- = −1
Oxidation state of Cl in ClO_3^- = +5
Oxidation state increase = oxidation
Oxidation state of S in SO_4^{2-} = +6
Oxidation state of S in SO_2 = +4
Oxidation state decrease = reduction
Oxidation and reduction have both occurred.
This *is* a redox reaction.

2.2 Electrochemical cells

2.2.1 Metal displacement reactions and galvanic cells

As mentioned previously, if the two half-equations that make up a metal displacement reaction are separated chemically but connected by a wire, it is possible to harness the energy possessed by the electrons to do work such as making a motor run, a bulb light up or a mobile phone operate.

This is achieved in the **galvanic cell** shown.

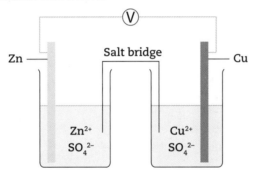

FIGURE 2.6 Galvanic cell between zinc and copper

Galvanic cell terminology

The zinc and copper strips are the electrodes.

The Zn strip is *negatively* charged and is called the **anode**. This is where **oxidation** takes place.

The copper strip is *positively* charged and is called the **cathode**. This is where **reduction** takes place.

The metal strips dip into solutions containing their respective ions:
Zn strip in $Zn^{2+}(aq)$
Cu strip in $Cu^{2+}(aq)$.

These solutions are termed the **electrolyte**. They are usually nitrate solutions because all NO_3^- salts are soluble.

The **salt bridge** is a link between the two half-cells which maintains the electrical neutrality of each half-cell. It can be a filter paper soaked in an electrolyte such as $NaNO_3(aq)$.

The strips of metal, solutions and salt bridge constitute the **internal circuit**, whereas the wires, voltmeter or appliance constitute the **external circuit**.

Referring to the net ionic equation for this reaction:

$$Zn(s) + Cu^{2+}(aq) \rightarrow Zn^{2+}(aq) + Cu(s)$$

At the anode, the Zn strip loses electrons, which flow out of the strip and into the external circuit.

These electrons eventually move into the copper strip in the cathode half-cell. The Cu^{2+} ions in the electrolyte are attracted to the negatively charged Cu strip and move towards it. When they hit the strip, they pick up electrons and become Cu atoms, which attach to the Cu strip.

2.2.2 Electrolytic cells

A galvanic cell is one in which a spontaneous redox reaction produces an electric current.

An **electrolytic cell** is one in which an electric current is used to force a non-spontaneous reaction to occur.

Consider the decomposition of molten lithium bromide:

$$2LiBr(l) \rightarrow 2Li(l) + Br_2(g)$$

The half-equations are:

$$Li^+(l) + e^- \rightarrow Li(l) \quad \text{reduction}$$

$$2Br^-(l) \rightarrow Br_2(g) + 2e^- \quad \text{oxidation}$$

Under normal conditions, this is a non-spontaneous reaction. The Li^+ and Br^- ions are far too stable.

However, they can be made to react by providing the electrons from an external power source.

Electrons flow from the power source, via the external circuit to the carbon electrode, making it negatively charged. The positively charged lithium ions in the molten electrolyte move towards the negative electrode, where they pick up an electron:

$$Li^+(l) + e^- \rightarrow Li(l)$$

This is reduction, so this electrode is the cathode.

As the Li^+ ions are reduced, more are attracted to the cathode, creating a flow of positive ions through the electrolyte towards the cathode.

The lithium metal produced melts because of the high temperature of the molten salt and forms a pool on the surface of the electrolyte around the cathode.

Completing the circuit is the other electrode, which is positively charged. Negatively charged bromide ions move towards this electrode. When they hit the electrode, they release their extra electrons:

$$2Br^-(l) \rightarrow Br_2(g) + 2e^-$$

This is oxidation, so this electrode is the anode.

Br^- ions are oxidised, so more are attracted to the anode, creating a flow of negative ions through the electrolyte towards the anode. The molecular bromine that is produced vaporises due to the high temperature of the molten salt and then dissipates away.

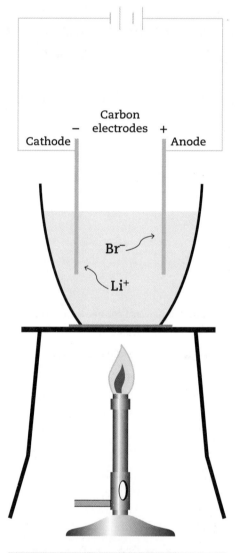

FIGURE 2.7 Apparatus used for the electrolysis of molten lithium bromide

2.2.3 Comparing galvanic and electrolytic cells

Galvanic cell	Electrolytic cell
Spontaneous reaction	Non-spontaneous reaction
Oxidation at anode	Reduction at cathode
Anode is negatively charged	Anode is positively charged
Electrons flow from anode to cathode	Electrons flow from anode to cathode
Positive ions flow in opposite direction	Positive ions flow in opposite direction

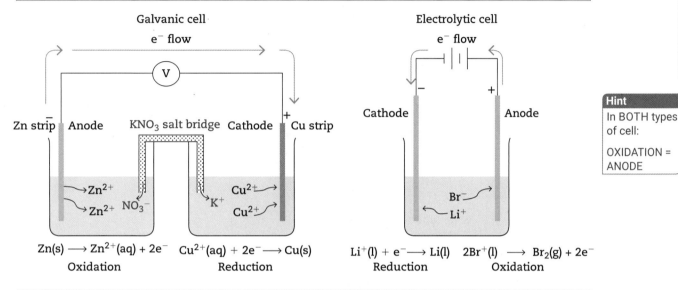

FIGURE 2.8 Diagram showing the similarities and differences between galvanic and electrolytic cells

Hint

In BOTH types of cell:

OXIDATION = ANODE

2.3 Galvanic cells

The spontaneous redox reaction

As mentioned previously, a galvanic cell uses a spontaneous redox chemical reaction to produce energy:

$$Zn(s) + Cu^{2+}(aq) \rightarrow Zn^{2+}(aq) + Cu(s)$$

Zinc and copper are both metals. They both want to lose electrons and be oxidised. However, Zn is more reactive than Cu and so will be the metal that is oxidised.

So, the following reaction is not spontaneous:

$$Cu(s) + Zn^{2+}(aq) \rightarrow Cu^{2+}(aq) + Zn(s)$$

Cu cannot, under normal circumstances, displace Zn^{2+} ions from solution.

So, in a galvanic cell using Zn and Cu metals, Zn is always the anode and Cu is always the cathode.

2.3.1 Fuel cells

Traditionally, hydrogen has been thought of as a fuel that releases its energy in the form of heat, when it is mixed with oxygen in a combustion reaction.

However, it is possible to extract energy directly, in the form of electrical energy, when mixing hydrogen with oxygen.

This is achieved in a **fuel cell**.

A fuel cell is a type of galvanic cell that continuously feeds in hydrogen gas and oxygen gas, releasing electrical energy when the hydrogen is oxidised and the oxygen is reduced.

Operation of an alkaline fuel cell

The alkaline fuel cell uses a potassium hydroxide electrolyte – hence the term *alkaline*. The operation of the fuel cell can best be described by the following steps.

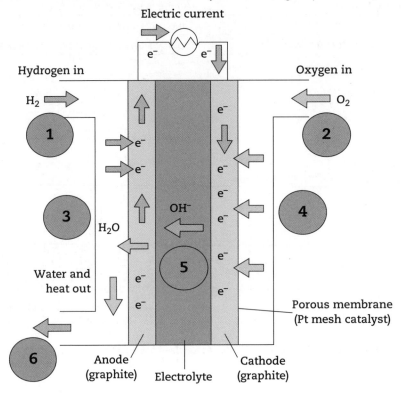

FIGURE 2.9 Alkaline fuel cell

Step 1: H_2 gas enters the cell.

Step 2: O_2 gas enters the cell.

Step 3: $H_2(g)$ passes through the porous platinum mesh that encloses the anode, where it is oxidised.

$$2H_2(g) + 4OH^-(aq) \rightarrow 4H_2O(l) + 4e^-$$

Step 4: The electrons produced move up and out of the electrode and into the external circuit, where they are used to power an appliance. The electrons then move down into the cathode.

Step 5: $O_2(g)$ passes through the porous platinum mesh that encloses the cathode, where it is reduced by the electrons produced in step 3.

$$O_2(g) + 2H_2O(l) + 4e^- \rightarrow 4OH^-(aq)$$

OH^- produced in step 4 passes through the porous platinum mesh and into the electrolyte. In the electrolyte, it migrates towards the anode to participate in the oxidation process described in step 3.

Step 6: The water and heat produced in this process leave the cell, and are recycled.

2.3.2 Inside the galvanic cell

In the galvanic cell in Figure 2.10, for the electrons released by the oxidation of Zn by Cu^{2+} ions to be used, the Zn and Cu^{2+} ions must be chemically separated. This results in two half-cells connected by a wire that conducts electrons and a salt bridge that allows the movement of ions.

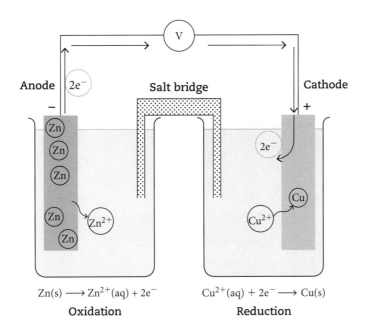

$$Zn(s) \longrightarrow Zn^{2+}(aq) + 2e^-$$
Oxidation

$$Cu^{2+}(aq) + 2e^- \longrightarrow Cu(s)$$
Reduction

FIGURE 2.10 The operation of a galvanic cell

Over time, the Zn electrode would appear to be smaller than it was at the beginning because much of it would have dissolved away during the oxidation process.

Meanwhile the Cu electrode would have become larger due to the build-up of fresh copper during the reduction process.

The salt bridge

It was mentioned earlier that the salt bridge can simply be a strip of filter paper dipped in an electrolyte solution such as $NaNO_3^-$.

It plays a critical role in the operation of the cell.

If the cell shown above was operating:

- In the anode half-cell, $Zn^{2+}(aq)$ ions are produced:

$$Zn(s) \rightarrow Zn^{2+}(aq) + 2e^-$$

This could cause a charge imbalance, but for every $Zn^{2+}(aq)$ ions produced, two $NO_3^-(aq)$ ions move down from the salt bridge to neutralise these positive ions.

- In the cathode half-cell, Cu^{2+} ions are being used up:

$$Cu^{2+}(aq) + 2e^- \rightarrow Cu(s)$$

Again, this could cause a charge imbalance, but for every $Cu^{2+}(aq)$ ion that disappears, two $Na^+(aq)$ ions move down from the salt bridge to replace the positive charge that has been lost.

2.3.3 The essential elements of a galvanic cell

A galvanic cell is based on a redox reaction such as a metal displacement reaction; for example:

$$Fe(s) + Cu(NO_3)_2(aq) \rightarrow Fe(NO_3)_2(aq) + Cu(s)$$

The Fe has been *oxidised*:

$$Fe(s) \rightarrow Fe^{2+}(aq) + 2e^-$$

The oxidation half-cell would consist of a beaker in which a strip of iron is dipping into a solution of $Fe^{2+}(aq)$ ions, such as $Fe(NO_3)_2(aq)$.

The $Cu^{2+}(aq)$ ions have been *reduced*:

$$Cu^{2+}(aq) + 2e^- \rightarrow Cu(s)$$

> **Hint**
> The oxidation state of a metal can be worked out by referring to the table of polyatomic ions on page 15 of the *Chemistry formula and data booklet*.

The reduction half-cell would consist of a beaker in which a strip of copper is dipping into a solution of Cu^{2+}(aq) ions, such as $Cu(NO_3)_2$(aq).

The two metal strips are the electrodes and their respective solutions are the *electrolytes*. Along with the salt bridge, they make up the internal circuit of the cell. The metal strips are connected to the wires and meters/appliances that make up the external circuit. Since the Fe strip is oxidised, it is the anode and because it is the source of electrons, it is the negative electrode. The Cu strip is the cathode and is the positive electrode.

Worked example

Question: The displacement reaction between magnesium and iron(II) nitrate is given by:

$$Mg(s) + Fe(NO_3)_2(aq) \rightarrow Mg(NO_3)_2(aq) + Fe(s)$$

Draw and label the galvanic cell represented by this equation.

Step 1

Remove any spectator ions and separate the equation into oxidation and reduction half-equations.

$$Mg(s) \rightarrow Mg^{2+}(aq) + 2e^- \quad \text{oxidation}$$

$$Fe^{2+}(aq) + 2e^- \rightarrow Fe(s) \quad \text{reduction}$$

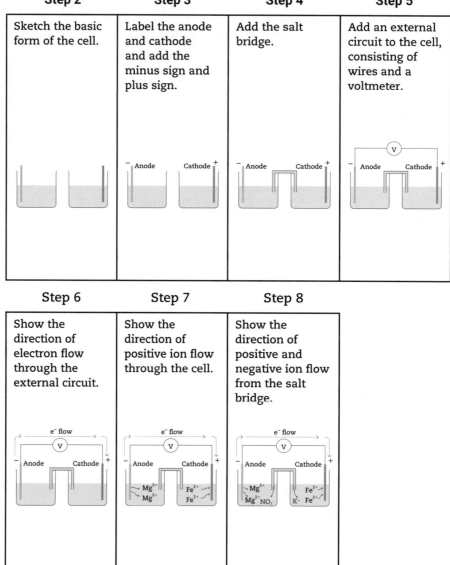

Step 2	Step 3	Step 4	Step 5
Sketch the basic form of the cell.	Label the anode and cathode and add the minus sign and plus sign.	Add the salt bridge.	Add an external circuit to the cell, consisting of wires and a voltmeter.

Step 6	Step 7	Step 8
Show the direction of electron flow through the external circuit.	Show the direction of positive ion flow through the cell.	Show the direction of positive and negative ion flow from the salt bridge.

Final diagram

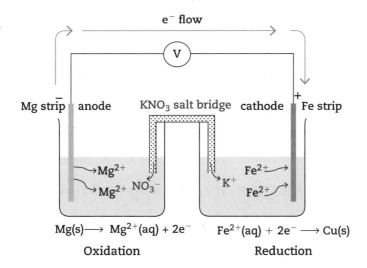

FIGURE 2.11 Worked example for a labelled galvanic cell represented by the equation
$Mg(s) + Fe(NO_3)_2(aq) \rightarrow Mg(NO_3)_2(aq) + Fe(s)$

2.3.4 Cell notation

Cell notation is a shorthand way of representing galvanic cells.

Consider the galvanic cell discussed in the worked example above. The cell notation for this cell would be:

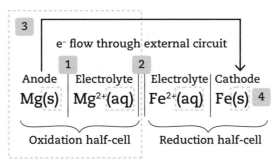

There are some important points to note.
1   A single vertical line represents a phase boundary between the solid electrode and the liquid electrolyte.
2   The double vertical line represents the salt bridge connecting the two half-cells.
3   The oxidation half-cell is always written first.
4   State symbols are essential.

Worked example

Question: Write the cell notation for a galvanic cell represented by the equation:

$$2Al(s) + 3Pb(NO_3)_2 \rightarrow 2Al(NO_3)_3(aq) + 3Pb(s)$$

Step 1

Remove any spectator ions and balancing coefficients, and separate the equation into oxidation and reduction half-equations.

$$Al(s) \rightarrow Al^{3+}(aq) + 3e^- \quad \text{oxidation}$$

$$Pb^{2+}(aq) + 2e^- \rightarrow Pb(s) \quad \text{reduction}$$

Hint
Note that the OXIDISED forms are closest to the salt bridge.

Step 2

Write the cell notation, remembering to put the oxidation half-cell first.

$$Al(s) \mid Al^{3+}(aq) \mid\mid Pb^{2+}(aq) \mid Pb(s)$$

2.4 Standard electrode potential

Standard electrode potential is a measure of the electrical potential of an electrochemical cell, under standard conditions. Standard electrode potential is represented by $E°$.

2.4.1 The relative strengths of oxidising and reducing agents

The potential of an electrode cannot be measured on its own, so it is important to be able to compare the relative oxidising and reducing strengths of substances. This can be achieved by comparing substances in a galvanic cell to a reference electrode.

The reference electrode used is the standard hydrogen electrode. This set-up is shown in Figure 2.12.

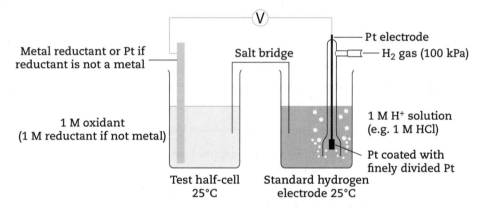

FIGURE 2.12 Set-up for determining the standard electrode potentials

2.4.2 Determining standard electrode potentials

All values are measured relative to the hydrogen half-cell which is given a value of 0.00 V. So, any reading on the voltmeter becomes the value of the test half-cell.

This method is used to produce a table of standard electrode potentials.

Standard electrode potentials at 25°C

Oxidised species ⇌ Reduced species	$E°$ (V)
$Li^+(aq) + e^- \rightleftharpoons Li(s)$	−3.04
$K^+(aq) + e^- \rightleftharpoons K(s)$	−2.94
$Ba^{2+}(aq) + 2e^- \rightleftharpoons Ba(s)$	−2.91
$Ca^{2+}(aq) + 2e^- \rightleftharpoons Ca(s)$	−2.87
$Na^+(aq) + e^- \rightleftharpoons Na(s)$	−2.71
$Mg^{2+}(aq) + 2e^- \rightleftharpoons Mg(s)$	−2.36
$Al^{3+}(aq) + 3e^- \rightleftharpoons Al(s)$	−1.68
$Mn^{2+}(aq) + 2e^- \rightleftharpoons Mn(s)$	−1.18

»

Oxidised species ⇌ Reduced species	$E°$ (V)
$2H_2O(l) + 2e^- \rightleftharpoons H_2(g) + 2OH^-(aq)$	−0.83
$Zn^{2+}(aq) + 2e^- \rightleftharpoons Zn(s)$	−0.76
$Fe^{2+}(aq) + 2e^- \rightleftharpoons Fe(s)$	−0.44
$Ni^{2+}(aq) + 2e^- \rightleftharpoons Ni(s)$	−0.24
$Sn^{2+}(aq) + 2e^- \rightleftharpoons Sn(s)$	−0.14
$Pb^{2+}(aq) + 2e^- \rightleftharpoons Pb(s)$	−0.13
$2H^+(aq) + 2e^- \rightleftharpoons H_2(g)$	0.00
$Cu^{2+}(aq) + e^- \rightleftharpoons Cu^+(aq)$	+0.16
$SO_4{}^{2-}(aq) + 4H^+(aq) + 2e^- \rightleftharpoons SO_2(aq) + 2H_2O(l)$	+0.16
$Cu^{2+}(aq) + 2e^- \rightleftharpoons Cu(s)$	+0.34
$O_2(g) + 2H_2O(l) + 4e^- \rightleftharpoons 4OH^-(aq)$	+0.40
$Cu^+(aq) + e^- \rightleftharpoons Cu(s)$	+0.52
$I_2(s) + 2e^- \rightleftharpoons 2I^-(aq)$	+0.54
$Fe^{3+}(aq) + e^- \rightleftharpoons Fe^{2+}(aq)$	+0.77
$Ag^+(aq) + e^- \rightleftharpoons Ag(s)$	+0.80
$Br_2(l) + 2e^- \rightleftharpoons 2Br^-(aq)$	+1.08
$O_2(g) + 4H^+(aq) + 4e^- \rightleftharpoons 2H_2O(l)$	+1.23
$Cl_2(g) + 2e^- \rightleftharpoons 2Cl^-(aq)$	+1.36
$Cr_2O_7{}^{2-}(aq) + 14H^+(aq) + 6e^- \rightleftharpoons 2Cr^{3+}(aq) + 7H_2O(l)$	+1.36
$MnO_4{}^-(aq) + 8H^+(aq) + 5e^- \rightleftharpoons Mn^{2+}(aq) + 4H_2O(l)$	+1.51
$F_2(g) + 2e^- \rightleftharpoons 2F^-(aq)$	+2.89

The reactive metals such as lithium, potassium and barium are towards the top of the table and have large negative values.

The reactive non-metals, such as fluorine and chlorine, are towards the bottom of the table and have large positive values.

The half-reactions are all written as reduction half-equations, which makes it easy to compare their relative strengths as oxidising or reducing agents.

When drawing a galvanic cell with, say, lead and nickel, it is possible to use the table to decide which metal is the anode and which is the cathode.

From the table, the half-equations as written are:

$$Ni^{2+}(aq) + 2e^- \rightleftharpoons Ni(s) \quad E° = -0.24 \text{ V}$$

$$Pb^{2+}(aq) + 2e^- \rightleftharpoons Pb(s) \quad E° = -0.13 \text{ V}$$

The nickel half-equation is higher in the table than the lead half-equation.
In a galvanic cell the anode is Ni, the cathode is Pb.

Hint

The higher in the table a metal is, the more reactive it is and the more easily it is oxidised.

9780170459150

2.4.3 Calculating standard cell potentials

The standard electrode potentials table can be used to determine the potential of half-cells in a galvanic cell, as well as the overall voltage supplied by the cell.

Consider a galvanic cell using nickel and lead electrodes.

As mentioned earlier, a strip of nickel is the anode and a strip of lead is the cathode.

To calculate the standard cell potential, $E°_{cell}$, the equation below can be used:

$$E°_{red} = E° \text{ (reduced species)} = Pb^{2+}(aq) + 2e^- \rightleftharpoons Pb(s) \quad E° = -0.13 \text{ V}$$

$$E°_{ox} = E° \text{ (oxidised species)} = Ni^{2+}(aq) + 2e^- \rightleftharpoons Ni(s) \quad E° = -0.24 \text{ V}$$

BUT as written in the table, the half-reaction for nickel is a reduction half-equation.

In the galvanic cell described, nickel is being oxidised, and so the half-equation must be reversed along with the sign of its voltage:

$$Ni(s) \rightarrow Ni^{2+}(aq) + 2e^- \quad E° = 0.24 \text{ V}$$

Therefore,

$$E°_{cell} = -0.13 + (+0.24) = +0.11 \text{ V}$$

Worked example

Question: Determine the standard cell potential, $E°_{cell}$, for a galvanic cell using aluminium and manganese electrodes.

Step 1

Identify the relevant equations from the table in section 2.4.2:

$$Al^{3+}(aq) + 3e^- \rightleftharpoons Al(s) \quad E° = -1.68 \text{ V}$$

$$Mn^{2+}(aq) + 2e^- \rightleftharpoons Mn(s) \quad E° = -1.18 \text{ V}$$

Step 2

Determine which metal is being oxidised and which is being reduced.

Al has a larger negative $E°$ value than Mn. Al is being oxidised and Mn^{2+} reduced.

Step 3

Use $E°_{cell} = E°_{red} + E°_{ox}$ to determine the cell potential, remembering to reverse the sign of the $E°_{ox}$ value.

$$E°_{cell} = -1.18 + (+1.68) = +0.5 \text{ V}$$

2.4.4 Standard electrode potentials – limitations

When using the table of standard cell potentials to determine the spontaneity of a redox equation and a galvanic cell's potential, some limitations must be considered.

Kinetic stability

In the event a cell potential for a reaction is calculated and found to have a positive value, it can be concluded that the reaction is spontaneous.

But these calculations do *not* consider factors such as the activation energy for a reaction.

For example, the reaction between magnesium and water proceeds according to the equation:

$$Mg(s) + H_2O(l) \rightarrow Mg^{2+}(aq) + 2OH^-(aq) + H_2(g)$$

The $E°_{cell}$ value for this reaction is +1.54 V. The activation energy for this reaction is high, which means, whilst the reaction *does* happen spontaneously, it occurs so slowly as to be imperceptible.

A high activation energy means that the reactants are **kinetically stable**.

Non-standard conditions

The table of standard electrode potentials is only useful if reactions are conducted under standard conditions of temperature (25°C), concentration (1 M) and pressure (gases must be at 100 kPa).

Other issues

It can be difficult to predict a cell voltage if the electrodes are very close together in the table.

Reactive metals, such as aluminium, which have already reacted with oxygen in the air to produce an impervious metal oxide coating, also present challenges.

2.4.5 Applications of standard electrode potentials

Calculations involving standard electrode potentials can come in three main categories.

1 Comparing relative strengths of oxidising and reducing agents

Worked example

Question: Arrange the list of substances below in order of decreasing reducing strength:

$$X (E° = +0.31 \text{ V}), D (E° = -2.32 \text{ V}), L (E° = -0.22 \text{ V}), T (E° = +0.54 \text{ V})$$

The strongest reducing agents are those substances most easily oxidised – the more reactive metals at the top of the table. The most reactive metals have the largest *negative E°* values.

Therefore:

$$D > L > X > T$$

2 Predicting whether a reaction will occur spontaneously

Worked example

Question: Predict whether tin metal will displace manganese metal from a solution of manganese(II) nitrate solution according to the equation:

$$Sn(s) + Mn(NO_3)_2(aq) \rightarrow Sn(NO_3)_2(aq) + Mn(s)$$

Step 1

Identify the relevant equations from the table of standard electrode potentials in section 2.4.2.

$$Mn^{2+}(aq) + 2e^- \rightleftharpoons Mn(s) \quad E° = -1.18 \text{ V}$$

$$Sn^{2+}(aq) + 2e^- \rightleftharpoons Sn(s) \quad E° = -0.14 \text{ V}$$

Step 2

By referring to the balanced equation, decide which metal is being oxidised and which is being reduced.

$$\text{Oxidation:} \quad Sn(s) \rightarrow Sn^{2+}(aq) + 2e^- \quad E° = -(-0.14) \text{ V}$$

$$\text{Reduction:} \quad Mn^{2+}(aq) + 2e^- \rightleftharpoons Mn(s) \quad E° = -1.18 \text{ V}$$

Step 3

Calculate $E°_{cell}$ using $E°_{cell} = E°_{red} + E°_{ox}$

$$E°_{cell} = -1.18 + (+0.14) = \mathbf{-1.04} \text{ V}$$

The negative $E°_{cell}$ value of −1.04 V indicates that this reaction is *not* spontaneous and so will *not* occur.

3 Calculating the cell potential of a galvanic cell

Worked example

Question: Calculate the standard cell potential for the cell containing silver and tin electrodes.

Step 1

Identify the relevant equations from the table of standard electrode potentials in section 2.4.2.

$$Sn^{2+}(aq) + 2e^- \rightleftharpoons Sn(s) \quad E° = -0.14 \text{ V}$$

$$Ag^+(aq) + e^- \rightleftharpoons Ag(s) \quad E° = +0.80 \text{ V}$$

Step 2

Decide which metal is the anode and which is the cathode by referring to the table.

The Sn half-equation has the larger *negative E°* value (it is closer to the top of the table).

Sn, therefore, is the element that is oxidised and Ag^+ is the ion that is reduced.

Therefore, Sn is the anode (oxidation) and Ag is the cathode (reduction).

Step 3

Calculate $E°_{cell}$ using $E°_{cell} = E°_{red} + E°_{ox}$

$$E°_{ox} = -(-0.14) \text{ V}$$

$$E°_{red} = +0.80 \text{ V}$$

$$E°_{cell} = +0.80 + (+0.14) = \mathbf{+0.94 \text{ V}}$$

2.5 Electrolytic cells

2.5.1 Electrolysis of molten salts

Electrolysis is defined as a chemical change caused by the application of an electric current.

For electrolysis to occur, electric current must be able to flow through the substance being oxidised.

To electrolyse a salt, such as magnesium chloride ($MgCl_2$), a great deal of heat must be applied.

$MgCl_2$ is an ionic substance and, as such, has a very high melting point. The salt needs to be converted into its liquid form so that the Mg^{2+} ions and Cl^- ions can move around and conduct an electric current.

Electrolytic cell operation

Step 1

Electrons move from the power pack and into the carbon electrode, causing it to be *negatively* charged.

Step 2

Positively charged molten Mg^{2+} ions move towards the *negatively* charged electrode.

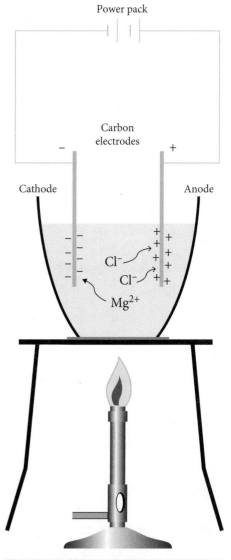

FIGURE 2.13 Set-up for electrolysis of molten magnesium chloride

Step 3

When electrons hit the electrode, the Mg^{2+} ions have electrons added to them:

$$Mg^{2+}(l) + 2e^- \rightarrow Mg(l)$$

This is reduction. Therefore, this is the CATHODE.

Step 4

The other electrode is *positively* charged. *Negatively* charged molten Cl^- ions move towards this electrode.

Step 5

When the Cl^- ions hit the electrode, they give up an electron each.

$$2Cl^-(l) \rightarrow Cl_2(g) + 2e^-$$

This is *oxidation*. Therefore, this is the ANODE.

If the cell potential for this electrolytic cell is calculated:

Reduction:	$Mg^{2+}(l) + 2e^- \rightarrow Mg(l)$	$E°_{red} = -2.36$ V
Oxidation:	$2Cl^-(l) \rightarrow Cl_2(g) + 2e^-$	$E°_{ox} = -(+1.36)$ V

$$E°_{cell} = -2.36 + (-1.36) = -3.72 \text{ V}$$

The negative sign for the $E°_{cell}$ indicates that this is *not* a spontaneous reaction.

Therefore, to force this reaction to occur, a minimum voltage of **3.72 V** must be applied from a power pack or a battery.

2.5.2 Electrolysis of aqueous solutions

These include many industrially and commercially important reactions.

Electroplating

Electroplating is an electrochemical process that enables a thin coating of a metal to be deposited on a metal object.

A very important application of this can be found in the cutlery industry where a metal such as silver can be applied to knives, forks and spoons to make them more durable and attractive.

Consider the electroplating of a metal fork with silver, as shown in Figure 2.14.

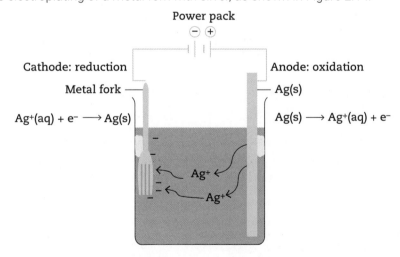

FIGURE 2.14 Set-up for silver-plating a metal fork

Hint

Remember, whether it is a galvanic cell or an electrolytic cell:

OXIDATION = ANODE

U3 – TOPIC 2

Electrorefining

This is very important in the purification of copper metal for use in electrical wiring. It is similar in many ways to electroplating but, instead, the anode is a piece of impure copper called 'blister' copper. The cathode is a small piece of 99.9 per cent pure copper. The electrolyte is a Cu^{2+} solution.

During the process the following occurs.

1 The copper from the impure anode is oxidised and goes into solution.

$$Cu(s) \rightarrow Cu^{2+}(aq) + 2e^-$$

2 The Cu^{2+} ions in solution move towards the *negatively* charged cathode.

3 When the Cu^{2+} ions hit the cathode, they are reduced.

$$Cu^{2+}(aq) + 2e^- \rightarrow Cu(s)$$

4 The $Cu(s)$ atoms coat the pure copper strip.

The process continues until all the copper has been removed from the impure 'blister' copper, leaving behind a block of 99.9 per cent copper.

2.5.3 Predicting and explaining products

Predicting the products of the electrolysis of molten binary salts, those that contain just two elements, is straightforward. They are the elements produced by the ions in the salt. Predicting the products of the electrolysis of molten polyatomic ions is beyond the scope of this course.

The electrolysis of aqueous solutions is complicated by the presence of water, which can also be electrolysed.

The two factors that need to be considered when predicting electrolysis products are the nature and concentration of the electrolyte.

The nature of the electrolyte

When electrolysing an aqueous solution, competing half-equations occur:

- Oxidation of electrolyte

- Oxidation of water

$$\text{Oxidation: } 2H_2O(l) \rightarrow O_2(g) + 4H^+(aq) + 4e^- \quad E° = -1.23 \text{ V}$$

- Reduction of electrolyte

- Reduction of water

$$\text{Reduction: } 2H_2O(l) + 2e^- \rightarrow H_2(g) + 2OH^-(aq) \quad E° = -0.83 \text{ V}$$

Worked example

Question: Predict the products of the electrolysis of a 1.00 M solution of copper(II) bromide, $Cu(Br)_2(aq)$.

Step 1

Given that the ions in solution are $Cu^{2+}(aq)$ and $Br^-(aq)$, use the table of standard electrode potentials on page 62 to write the oxidation and reduction half-equations:

$$\text{Oxidation: } 2Br^-(aq) \rightarrow Br_2(l) + 2e^- \quad E° = -1.08 \text{ V}$$

$$\text{Reduction: } Cu^{2+}(aq) + 2e^- \rightarrow Cu(s) \quad E° = +0.34 \text{ V}$$

Step 2

Identify the competing reactions at the anode and cathode.

Anode (oxidation):

$$2Br^-(aq) \rightarrow Br_2(l) + 2e^- \quad\quad E° = \mathbf{-1.08} \text{ V}$$

$$2H_2O(l) \rightarrow O_2(g) + 4H^+(aq) + 4e^- \quad E° = \mathbf{-1.23} \text{ V}$$

Cathode (reduction):

$$Cu^{2+}(aq) + 2e^- \rightarrow Cu(s) \qquad E° = +0.34 \text{ V}$$

$$2H_2O(l) + 2e^- \rightarrow H_2(g) + 2OH^-(aq) \quad E° = -0.83 \text{ V}$$

Step 3

Predict the products.

Anode:

$E° = -1.08$ (Br$_2$) is less negative than $E° = -1.23$ (H$_2$O)

Therefore, **Br$_2$(l)** is produced at the anode.

Cathode:

$E° = +0.34$ (Cu) is less negative than $E° = -0.83$ (H$_2$O)

Therefore, **Cu(s)** is produced at the cathode.

Worked example

Question: Predict the products of the electrolysis of a 1.00 M solution of sodium chloride, NaCl(aq).

Step 1

Given that the ions in solution are Na$^+$(aq) and Cl$^-$(aq), use the table of standard electrode potentials on page 62 to write the oxidation and reduction half-equations.

$$\text{Oxidation: } 2Cl^-(aq) \rightarrow Cl_2(l) + 2e^- \qquad E° = -1.36 \text{ V}$$

$$\text{Reduction: } Na^+(aq) + e^- \rightarrow Na(s) \qquad E° = -2.71 \text{ V}$$

Remember, the oxidation and reduction half-equations for water can be challenging. It is much easier to work them out from first principles.

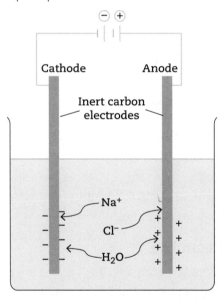

FIGURE 2.15 Set-up for the electrolysis of a 1.00 M NaCl solution

Consider the reduction of water at the cathode. Normally, it would be Na$^+$, the *positive* ion, that is reduced.

What part of H$_2$O would be considered the *positive* part? The answer should be H$^+$.

When H$_2$O hits the cathode, the H$^+$ in the H$_2$O is reduced to hydrogen, H$_2$.

So,

$$H_2O(l) + e^- \rightarrow H_2(g)$$

The only half-equation in the table of standard electrode potentials on page 62 similar to this arrangement is:

$$2H_2O(l) + 2e^- \rightarrow H_2(g) + 2OH^-(aq) \quad E° = -0.83 \text{ V}$$

U3 – TOPIC 2

Consider the oxidation of water at the anode. Normally, it would be Cl^-, the *negative* ion, that is oxidised.

What part of H_2O would be considered the *negative* part? The answer should be O^{2-}.

When H_2O hits the cathode, the O^{2-} in the H_2O is reduced to oxygen, O_2.

So,

$$H_2O(l) \rightarrow O_2(g) + e^-$$

The only half-equation in the table on page 62 that is similar to this arrangement is:

$$2H_2O(l) \rightarrow O_2(g) + 4H^+(aq) + 4e^- \quad E° = -1.23 \text{ V}$$

Now, we can move on to step 2 of this worked example.

Step 2

Identify the possible anode and cathode reactions.

Anode (oxidation):

$$2Cl^-(aq) \rightarrow Cl_2(l) + 2e^- \qquad E° = -1.36 \text{ V}$$

$$2H_2O(l) \rightarrow O_2(g) + 4H^+(aq) + 4e^- \qquad E° = -1.23 \text{ V}$$

Cathode (reduction):

$$Na^+(aq) + e^- \rightarrow Na(s) \qquad E° = -2.71 \text{ V}$$

$$2H_2O(l) + 2e^- \rightarrow H_2(g) + 2OH^-(aq) \quad E° = -0.83 \text{ V}$$

Step 3

Predict the products.

Anode:

$E° = \textbf{–1.23}$ (H_2O) is less negative than $E° = \textbf{–1.36}$ (Cl_2)

Therefore, **$O_2(g)$** is produced at the anode.

Cathode:

$E° = \textbf{–0.83}$ (H_2O) is less negative than $E° = \textbf{–2.71}$ (Na)

Therefore, **$H_2(g)$** is produced at the cathode.

Electrolysis of aqueous polyatomic ions

Polyatomic anions, such as sulfate (SO_4^{2-}), nitrate (NO_3^-) and phosphate (PO_4^{3-}), are much more stable than monatomic anions. Hence, much larger voltages are required for them to be oxidised. Thus, when polyatomic anions are present in solution, the water will always be oxidised preferentially to the polyatomic anions because it has a less negative voltage.

The concentration of the electrolyte

When predicting the products of aqueous solutions, the rule that the reaction with the lowest negative voltage is the favoured reaction is always true.

Issues can arise when the competing reactions have similar $E°$ values. In situations like this, the *concentration* of the electrolyte becomes important.

Consider the electrolysis of a solution of $CuCl_2$, the product at the cathode is obvious.

Cathode (reduction):

$$Cu^{2+}(aq) + 2e^- \rightarrow Cu(s) \qquad E° = +0.34 \text{ V}$$

$$2H_2O(l) + 2e^- \rightarrow H_2(g) + 2OH^-(aq) \quad E° = -0.83 \text{ V}$$

Cu(s) is produced. Its reaction is much less negative than that for H_2O.

But at the
Anode (oxidation):

$$2Cl^-(aq) \rightarrow Cl_2(l) + 2e^- \qquad E° = \mathbf{-1.36}\ V$$

$$2H_2O(l) \rightarrow O_2(g) + 4H^+(aq) + 4e^- \qquad E° = \mathbf{-1.23}\ V$$

These values are quite similar so, in this case, concentration has an effect.
In a concentrated solution, the oxidation of Cl^- ions is more likely (there are more of them).
In a dilute solution, oxidation of H_2O is more likely.

Worked example

Question: Predict the products of the electrolysis of a 4.00 M solution of sodium chloride, NaCl(aq).
Write an equation for the overall reaction and determine the minimum voltage required to enable this
reaction to occur.

Step 1

Identify the relevant half-equations.
   Cathode (reduction):

$$Na^+(aq) + e^- \rightarrow Na(s) \qquad E° = -2.71\ V$$

OR

$$2H_2O(l) + 2e^- \rightarrow H_2(g) + 2OH^-(aq) \qquad E° = -0.83\ V$$

   Anode (oxidation):

$$2Cl^-(aq) \rightarrow Cl_2(l) + 2e^- \qquad E° = -1.36\ V$$

OR

$$2H_2O(l) \rightarrow O_2(g) + 4H^+(aq) + 4e^- \qquad E° = -1.23\ V$$

Step 2

Use the rule that reactions with the *least negative* voltages are preferred to determine which
half-reactions occur at each electrode:
   Cathode (reduction):

$$2H_2O(l) + 2e^- \rightarrow H_2(g) + 2OH^-(aq) \qquad E° = -0.83\ V$$

Therefore, **H_2(g)** is produced at the cathode.
Anode (oxidation):

$$2H_2O(l) \rightarrow O_2(g) + 4H^+(aq) + 4e^- \qquad E° = -1.23\ V$$

O_2(g) *should* be produced but the concentration is 4.00 M – concentrated.
Therefore, Cl^- ions are oxidised.

$$2Cl^-(aq) \rightarrow Cl_2(g) + 2e^- \qquad E° = -1.36\ V$$

Therefore, **Cl_2(g)** is produced at the cathode.

Step 3

Add the half-equations to give the overall equation, remembering to cancel the $2e^-$.

$$2H_2O(l) + 2Cl^-(aq) \rightarrow H_2(g) + 2OH^-(aq) + Cl_2(g)$$

Step 4

Use $E°_{cell} = E°_{red} + E°_{ox}$ to determine the voltage required.

$$E°_{cell} = -0.83 + (-1.36) = \mathbf{-2.19}\ V$$

Glossary

anode
the negatively charged electrode, at which oxidation takes place

cathode
the positively charged electrode, at which reduction takes place

electrolysis
a chemical change caused by the application of an electric current

electrolyte
a substance that conducts an electric current

electrolytic cell
a cell in which an electric current is used to force a non-spontaneous reaction to occur

external circuit
the wires, voltmeter or appliance in a galvanic cell

fuel cell
a type of galvanic cell that continuously feeds in hydrogen gas and oxygen gas, releasing electrical energy when the hydrogen is oxidised, and the oxygen is reduced

galvanic cell
a cell in which a spontaneous redox reaction produces an electric current

internal circuit
the strips of metal, solutions and salt bridge in a galvanic cell

kinetically stable
describes a cell where the cell potential indicates that a reaction will occur but, in practice, there will be no perceptible reaction because of the high activation energy

oxidant or oxidising agent
a chemical species that causes another species to be oxidised

oxidation
the loss of electrons

oxidation state
the number of electrons that an atom would have gained or lost to get to its present state; it assumes that each species is completely ionic

redox or oxidation–reduction reactions
a reaction that involves the oxidation of one species and the reduction of another species. The 'red-' comes from the 'reduction' half-equation and the '-ox' comes from 'oxidation' half-equation

reductant or reducing agent
a chemical species that causes another to be reduced

reduction
the gain of electrons

salt bridge
a link containing electrolyte between the oxidation and reduction half cells of a galvanic cell

standard electrode potential
a measure of the electrical potential of an electrochemical cell, under standard conditions. Standard electrode potential is represented by $E°$

9780170459150

Revision summary

Use the following summary of syllabus dot points and key knowledge within Unit 3 Topic 2 to ensure that you have thoroughly reviewed the content. Provide a brief definition or comment for each item to demonstrate your understanding or code them using the traffic light system: Green (all good); Amber (needs some review); Red (priority area to review).

Redox reactions	
• recognise that a range of reactions, including displacement reactions of metals, combustion, corrosion and electrochemical processes, can be modelled as redox reactions involving oxidation of one substance and reduction of another substance	
• understand that the ability of an atom to gain or lose electrons can be predicted from the atom's position in the periodic table, and explained with reference to valence electrons, consideration of energy and the overall stability of the atom	
• identify the species oxidised and reduced, and the oxidising agent and reducing agent, in redox reactions	
• understand that oxidation can be modelled as the loss of electrons from a chemical species, and reduction can be modelled as the gain of electrons by a chemical species; these processes can be represented using balanced half-equations and redox equations (acidic conditions only)	››

››	• deduce the oxidation state of an atom in an ion or compound and name transitional metal compounds from a given formula by applying oxidation numbers represented as roman numerals	
	• use appropriate representations, including half-equations and oxidation numbers, to communicate conceptual understanding, solve problems and make predictions	
	• Mandatory practical: Perform single displacement reactions in aqueous solutions.	
	Electrochemical cells	
	• understand that electrochemical cells, including galvanic and electrolytic cells, consist of oxidation and reduction half-reactions connected via an external circuit that allows electrons to move from the anode (oxidation reaction) to the cathode (reduction reaction)	
	Galvanic cells	
	• understand that galvanic cells, including fuel cells, generate an electrical potential difference from a spontaneous redox reaction, which can be represented as cell diagrams including anode and cathode half-equations	››

»

• recognise that oxidation occurs at the negative electrode (anode) and reduction occurs at the positive electrode (cathode) and explain how two half-cells can be connected by a salt bridge to create a voltaic cell (examples of half-cells are Mg, Zn, Fe and Cu and their solutions of ions)	
• describe, using a diagram, the essential components of a galvanic cell; including the oxidation and reduction half-cells, the positive and negative electrodes and their solutions of their ions, the flow of electrons and the movement of ions, and the salt bridge	
• Mandatory practical: Construct a galvanic cell using two metal/metal-ion half cells.	

Standard electrode potential

• determine the relative strength of oxidising and reducing agents by comparing standard electrode potentials	
• recognise that cell potentials at standard conditions can be calculated from standard electrode potentials; these values can be used to compare cells constructed from different materials	
• recognise the limitation associated with standard reduction potentials	

»

» · use appropriate mathematical representation to solve problems and make predictions about spontaneous reactions, including calculating cell potentials under standard conditions	
Electrolytic cells	
· understand that electrolytic cells use an external electrical potential difference to provide the energy to allow a non-spontaneous redox reaction to occur, and appreciate that these can be used in small-scale and industrial situations, including metal plating and the purification of copper	
· predict and explain the products of the electrolysis of a molten salt and aqueous solutions of sodium chloride and copper sulfate. Explanations should refer to E^o values, the nature of the electrolyte and the concentration of the electrolyte	
· describe, using a diagram, the essential components of an electrolytic cell; including source of electric current and conductors, positive and negative electrodes, and the electrolyte.	

Chemistry General Senior Syllabus 2019, © State of Queensland (QCAA) 2019, licensed under CC BY 4.0

Exam practice

Topic 2: Oxidation and reduction

Multiple-choice questions

Each multiple-choice question is worth 1 mark.

Solutions start on page 171.

Question 1

Which one of the following equations represents a redox reaction?

A $Pb(NO_3)_2(aq) + 2NaBr(aq) \rightarrow 2NaNO_3(aq) + PbBr_2(s)$

B $2NaOH(aq) + H_2SO_4(aq) \rightarrow Na_2SO_4(aq) + H_2O(l)$

C $Pb(NO_3)_2(aq) + Mn(s) \rightarrow Mn(NO_3)_2(aq) + Pb(s)$

D $MgCO_3(s) \rightarrow MgO(s) + CO_2(g)$

Question 2

Which one of the following is true regarding electrolytic and galvanic cells?

	Electrolytic cell	Galvanic cell
A	Spontaneous reaction	Oxidation at anode
B	Oxidation at cathode	Electrons flow from anode to cathode
C	Non-spontaneous reaction	Cathode is negatively charged
D	Anode is positively charged	Electrons flow from anode to cathode

Question 3

What is the role of the salt bridge in a galvanic cell?

A Maintains electrical neutrality in the cell

B Allows electrons to pass from one cell to another

C Allows ions to pass from the anode half-cell to the cathode half-cell

D Enables the solutions in each half-cell to mix

Question 4

Calculate the cell potential produced by a galvanic cell represented by the following equation:

$$Fe(s) + Pb(NO_3)_2(aq) \rightarrow Fe(NO_3)_2(aq) + Pb(s)$$

A +0.57 V

B +0.31 V

C −0.75 V

D −0.31 V

Question 5

The diagram below represents the apparatus used for the electrorefining of copper.

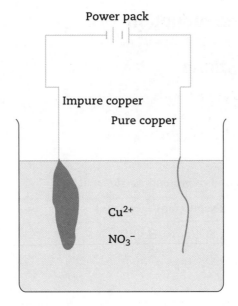

Which one of the following statements is true?

A The impure copper is the cathode, which dissolves.

B The pure copper is the cathode, and it produces electrons.

C The pure copper is negatively charged and attracts positively charged ions.

D Electrons flow from the power pack and reduce the copper in the impure copper.

Question 6

Deduce the oxidation state of iodine in the iodate ion, IO_3^-.

A +6   C +7

B +5   D +4

Question 7

Calculate the minimum voltage required to electrolyse a concentrated solution of copper(II) chloride, $CuCl_2$.

A −1.02 V   C +1.02 V

B +1.07 V   D −2.38 V

Question 8

The reaction between permanganate, MnO_4^-, and oxalate, $5C_2O_4^{2-}$, is given by the equation:

$$2MnO_4^-(aq) + 16H^+(aq) + 5C_2O_4^{2-}(aq) \rightarrow 2Mn^{2+}(aq) + 10CO_2(aq) + 8H_2O(l)$$

Which one of the following statements about this reaction is true?

A Oxalate is the oxidising agent because it gains electrons.

B Permanganate is the oxidising agent because it loses electrons.

C Oxalate is the reducing agent because it loses electrons.

D Permanganate is the oxidising agent because the oxidation state of Mn increases.

Question 9

In a galvanic cell, cell voltage is

A positive due to a spontaneous chemical reaction.

B negative due to a non-spontaneous chemical reaction.

C positive due to a non-spontaneous chemical reaction.

D negative due to a spontaneous chemical reaction.

Question 10

Which one of the following is true for the electrolytic cell shown?

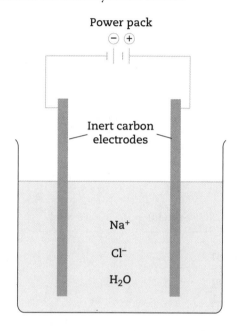

	Product at anode	Product at cathode
A	Hydrogen	Oxygen
B	Oxygen	Hydrogen
C	Sodium	Chlorine
D	Chlorine	Hydrogen

Question 11

Identify the correct order of the strength of the reducing agents by comparing their standard electrode potentials.

A $Ag > Cl^- > Pb$

B $Cl^- > Pb > Ag$

C $Pb > Ag > Cl^-$

D $Cl^- > Ag > Pb$

Question 12

The combustion of ethane is given by the equation:

$$C_3H_8(g) + O_2(g) \rightarrow CO_2(g) + H_2O(l)$$

This is classified as a redox reaction because the oxidation state of

A carbon increases and the oxidation state for oxygen decreases.

B carbon decreases and the oxidation state for hydrogen increases.

C carbon decreases and the oxidation state for oxygen increases.

D hydrogen decreases and the oxidation state for oxygen increases.

Question 13

Identify the correct shorthand cell notation for the galvanic cell shown below.

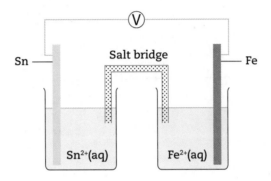

A $Sn(s) \mid Sn^{2+}(aq) \parallel Fe^{2+}(aq) \mid Fe(s)$

B $Fe^{2+}(aq) \mid Fe(s) \parallel Sn^{2+}(aq) \mid Sn(s)$

C $Fe(s) \mid Fe^{2+}(aq) \parallel Sn(s) \mid Sn^{2+}(aq)$

D $Fe(s) \mid Fe^{2+}(aq) \parallel Sn^{2+}(aq) \mid Sn(s)$

Question 14

The half-equations representing the corrosion of an iron object are:

$$Fe(s) \rightarrow Fe^{2+} + 2e^-$$

$$O_2(g) + 2H_2O(l) \rightarrow 4OH^-(aq)$$

Which one of the following statements regarding deep ocean shipwrecks is correct?

A Corrosion occurs very slowly because it is very cold.

B Corrosion occurs very slowly because of the high pressure of water.

C Corrosion occurs very slowly because there is very little dissolved oxygen.

D Corrosion does not occur because the iron gets coated with algae.

Short response questions

Solutions start on page 172.

Question 15 (4 marks)

The oxidation of ethanol by potassium dichromate crystals used to be the basis behind the operation of breathalysers used by police to detect alcohol on the breath of motorists.

The partial half-equations for this reaction are:

$$Cr_2O_7^{2-}(aq) \rightarrow Cr^{3+}(aq)$$

$$C_2H_5OH(aq) \rightarrow CH_3COOH(aq)$$

Apply your understanding of redox reactions to write the balanced equation for this reaction in acidic conditions.

Question 16 (7 marks)

A student was performing single displacement reactions in the lab. They placed four unknown metals, Q, T, L and D, into wells on a spotting plate containing solutions of their respective nitrates.

Below is a copy of the student's laboratory notebook.

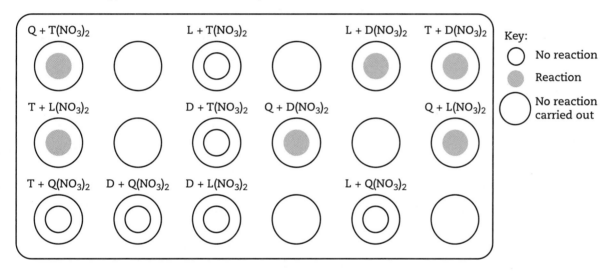

a Identify the reductant in the reaction between L and D(NO$_3$)$_2$.   1 mark

b Determine the oxidation state of metal Q in its nitrate solution.   1 mark

c Predict the order of reducing agent strength by matching metals Q, T, L and D with their respective standard reduction potentials in the table below. Include the reduction half-equation for each metal. Explain your reasoning.   5 marks

Oxidised species ⇌ Reduced species	$E°$ (V)
	−0.24
	−0.13
	+0.14
	+0.73

Question 17 (8 marks)

Below is a diagram of an alkaline fuel cell.

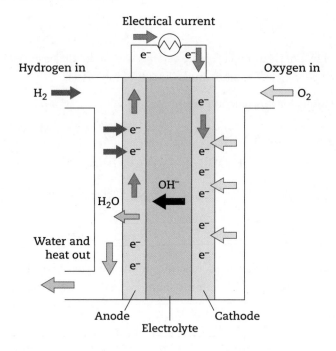

a Identify the products at the anode and cathode by writing the appropriate half-equations.   4 marks

   Anode half-equation

   Anode product

   Cathode half-equation

   Cathode product

b With reference to the table of standard electrode potentials, determine the voltage
   produced by the fuel cell.   3 marks

c Determine the overall reaction for the fuel cell.   1 mark

Question 18 (7 marks)

Below is a diagram of an electrolytic cell used for the industrial production of sodium.

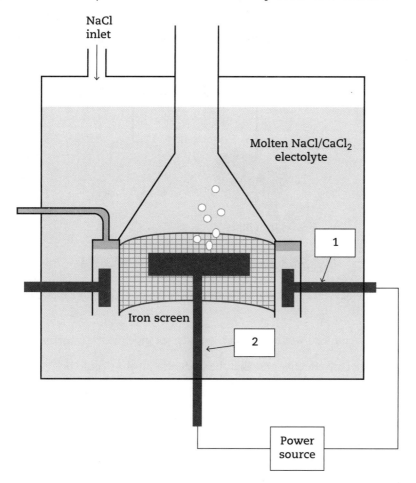

a   Identify the parts of the cell labelled 1 and 2.   2 marks

b   Write the oxidation and reduction half-equations for this process.   2 marks

c   Determine the minimum voltage required to run this cell.   3 marks

Question 19 (5 marks)

The diagram below represents a galvanic cell.

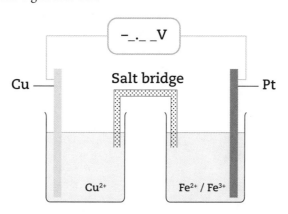

a   Identify the anode and cathode half-equations.   2 marks

b   Determine the net ionic equation for this cell.   1 mark

c   Predict the reading on the voltmeter.   2 marks

Question 20 (6 marks)

Corrosion of iron and steel structures, such as buildings, ships and undersea pipelines, is an ever-present problem, and companies spend large amounts of money to protect against this damage.

A cost-effective method of corrosion protection is to connect metal blocks to the structure, as shown in the diagram below.

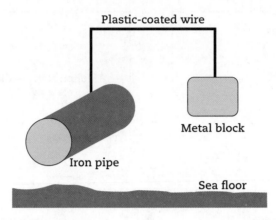

a Apply your knowledge of standard electrode potentials to suggest a suitable metal for the block.   2 marks

b Explain how your chosen metal would protect the undersea pipeline from corrosion.   2 marks

c Describe an alternative cost-effective way that the undersea pipeline could be protected against corrosion.   2 marks

Chapter 3
Unit 3 Data test

The Data test is the first of the summative internal assessments and is specifically associated with Unit 3 content.

It addresses Assessment Objectives 2, 3 and 4, requiring you to apply understanding, and analyse and interpret evidence. It is completed individually, under supervised conditions, with 60 minutes of working time and 10 minutes of perusal time.

TABLE 3.1 Summary of the types of responses possible in IA1

	The response must be:	Verbs:
Apply understanding	An unknown scientific quantity or feature	Calculate (show your working), identify, recognise, use evidence
Analyse evidence	A trend, pattern, relationship, limitation or uncertainty in the data sets	Categorise, classify, contrast, distinguish, organise, sequence
Interpret evidence	A conclusion based on the data sets (not your general knowledge)	Compare, deduce, extrapolate, infer, justify, predict

It is important to match the type of response to the verb in the question.

TABLE 3.2 Definitions of cognitive verbs associated with the data test assessment objectives

Assessment objective	Verb	Definition (QCAA)
Apply	Calculate	Determine or find (e.g. a number, answer) by using mathematical processes; obtain a numerical answer showing the relevant stages of working; ascertain/determine from given facts, figures or information.
	Identify	Distinguish; locate, recognise and name; establish or indicate who or what someone or something is; provide an answer from a number of possibilities; recognise and state a distinguishing fact or figure.
	Recognise	Identify or recall particular features of information from knowledge; identify that an item, characteristic or quality exists; perceive as existing or true; be aware of or acknowledge.
	Use evidence	Operate or put into effect; apply knowledge or rules to put theory into practice.
Analyse	Categorise	Place in or assign to a particular class or group; arrange or order by classes or categories; classify, sort out, sort, separate.
	Classify	Arrange, distribute or order in classes or categories according to shared qualities or characteristics.
	Contrast	Display recognition of differences by deliberate juxtaposition of contrary elements; show how things are different or opposite; give an account of the differences between two or more items or situations, referring to both or all of them throughout.
	Distinguish	Recognise as distinct or different; note points of difference between; discriminate; discern; make clear difference/s between two or more concepts or items.
	Organise	Arrange, order; form as or into a whole consisting of interdependent or coordinated parts, especially for harmonious or united action.
	Sequence	Place in a continuous or connected series; arrange in a particular order.

»

Assessment objective	Verb	Definition (QCAA)
Interpret	Compare	Display recognition of similarities and differences and recognise the significance of these similarities and differences.
	Deduce	Reach a conclusion that is necessarily true, provided a given set of assumptions is true; arrive at, reach or draw a logical conclusion from reasoning and the information given.
	Extrapolate	Infer or estimate by extending or projecting known information; conjecture; infer from what is known; extend the application of something (e.g. a method or conclusion) to an unknown situation by assuming that existing trends will continue or similar methods will be applicable.
	Infer	Derive or conclude something from evidence and reasoning, rather than from explicit statements; listen or read beyond what has been literally expressed; imply or hint at.
	Justify	Give reasons or evidence to support an answer, response or conclusion; show or prove how an argument, statement or conclusion is right or reasonable.
	Predict	Give an expected result of an upcoming action or event; suggest what may happen based on available information.

Chemistry General Senior Syllabus 2019, © State of Queensland (QCAA) 2019, licensed under CC BY 4.0

Conditions

- Time: 60 minutes plus 10 minutes perusal.
- Length: 400–500 words in total, consisting of
  - Short responses, i.e. sentence or short paragraphs
  - Written paragraphs, 50–250 words per item
- Other types of item responses, e.g. interpreting and calculating, should allow students to complete the response in the set time.
- Other:
  - Queensland-approved graphics calculator permitted
  - Data booklet permitted
  - Unseen stimulus may be included.

Data set 1

Data collected from the reaction between sulfur dioxide and oxygen in an enclosed system is shown in Figure 3.1.

$$SO_2(g) + O_2(g) \rightleftharpoons SO_3(g) \quad \Delta H = -196 \text{ kJ mol}^{-1}$$

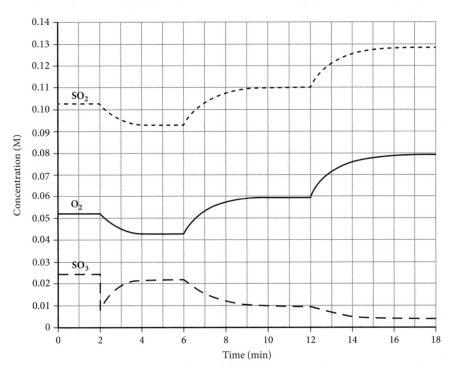

FIGURE 3.1 Concentration versus time graph for the $SO_2(g) + O_2(g) \rightleftharpoons SO_3(g)$ equilibrium system

Question 1 (2 marks)

Compare the relative quantities of SO_2 and SO_3 at 5 minutes and 17 minutes.

Question 2 (3 marks)

Distinguish the changes that were made to the system at 2 minutes and 12 minutes.

Question 3 (5 marks)

Using data from the graphs at 11 minutes and 17 minutes, write a statement relating the magnitude of the equilibrium constant, K_c, and temperature.

Data set 2

An experiment was carried out to determine the concentration and identity of an unknown monoprotic acid. A sample of the acid was diluted by a factor of 20. A 20.00 mL sample of this was titrated with a 0.103 M solution of NaOH. The progress of the reaction was monitored with a pH probe. The results are shown in Figure 3.2.

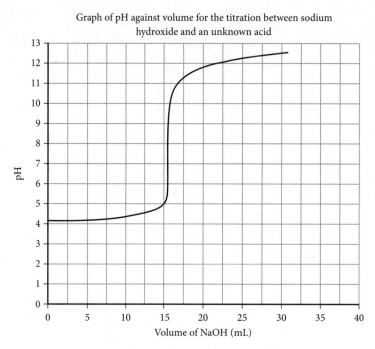

FIGURE 3.2 pH curve of a titration between sodium hydroxide and an unknown acid

Question 1 (1 mark)

Determine the pH of the equivalence point. Give your answer correct to one decimal place.

Question 2 (3 marks)

Distinguish the half-equivalence point from the equivalence point with reference to pH and volume of NaOH added.

Question 3 (3 marks)

With reference to the titration curve and using the table below, identify the acid used in the titration.

Acid	K_a
Ethanoic	1.76×10^{-5}
Propanoic	1.34×10^{-5}
Butanoic	1.54×10^{-5}
Heptanoic	7.10×10^{-6}
Decanoic	1.43×10^{-5}
Benzoic	6.46×10^{-5}
Chlorobenzoic	1.04×10^{-4}

Question 4 (3 mark)

Calculate the concentration of the acid solution in the 20.00 mL sample. Give your answer to two decimal places.

UNIT 4
STRUCTURE, SYNTHESIS AND DESIGN

Chapter 4
Topic 1: Properties and structure of organic materials

Topic summary

Organic chemistry, the chemistry of carbon-based molecules, is critical to many applications of chemistry.

In this topic, the **nomenclature** and terminology used to describe and distinguish organic substances is studied. This allows patterns and trends within and between different classes of organic substances to be observed and explained.

In addition, the specific types of reaction that each class of organic molecule participates in are studied, to construct synthetic reaction pathways from a series of reactions.

These characteristics and reaction pathways are incorporated into understanding and explaining the behaviour of biomolecules.

Finally, the development of machines and instruments used to analyse and determine the structure of complex organic and biomolecules is studied, along with ways to interpret the information that these instruments provide.

9780170459150

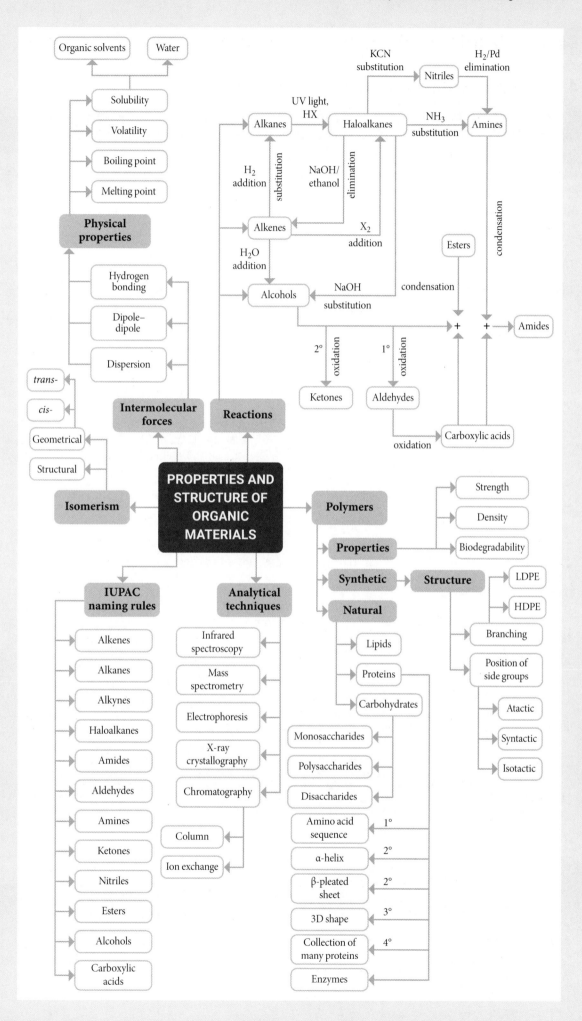

4.1 Structure of organic compounds

4.1.1 Representing organic compounds

The reason why organic chemistry is the chemistry of carbon and encompasses literally millions of compounds is the ability of carbon to **catenate**, which means it forms bonds with other carbon atoms to form long chains and rings, known as the **carbon skeleton.**

Organic compounds are grouped into series of families, called **homologous series**. The members of a homologous series have the same functional group and similar chemical properties. However, their carbon chains differ in length, thus their physical properties can vary accordingly.

An example of a homologous series is the **alkane** series in Table 4.1.

TABLE 4.1 Names of the first ten alkanes

Number of carbon atoms	Alkane name stem	Alkane name
1	Meth-	Methane (CH_4)
2	Eth-	Ethane (C_2H_6)
3	Prop-	Propane (C_3H_8)
4	But-	Butane (C_4H_{10})
5	Pent-	Pentane (C_5H_{12})
6	Hex-	Hexane (C_6H_{14})
7	Hept-	Heptane (C_7H_{16})
8	Oct-	Octane (C_8H_{18})
9	Non-	Nonane (C_9H_{20})
10	Dec-	Decane ($C_{10}H_{22}$)

Note that each member of the series differs from the ones next to it by $-CH_2$.

The other feature of organic compounds is the **functional group**. The functional group is a group of atoms in a molecule that gives the molecule distinctive chemical and physical properties.

TABLE 4.2 Functional groups

Name	Functional group	Prefix or suffix	Homologous series	Example	Name of example
Alkene	$-C=C-$	-ene	Alkenes		Ethene
Alkyne	$-C\equiv C-$	-yne	Alkyne		Propyne
Haloalkane	Halogen $-F, -Cl, -Br, -I$	Fluoro-, chloro-, bromo-, iodo-	Haloalkanes		Bromoethane
Hydroxyl	$-OH$	-anol or hydroxyl-	Alcohols		Ethanol
Aldehyde	$\overset{O}{\overset{\|}{-C-H}}$	-anal	Aldehydes		Propanal

Name	Functional group	Prefix or suffix	Homologous series	Example	Name of example
Ketone	O‖ —C—	-anone or oxo-	Ketones	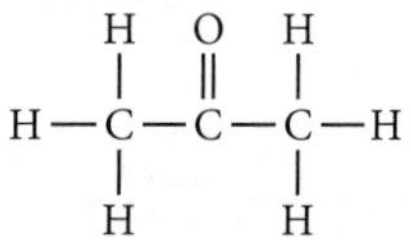 H—C(H)(H)—C(=O)—C(H)(H)—H	Propanone
Carboxyl	O‖ —C—OH	-anoic acid	Carboxylic acids	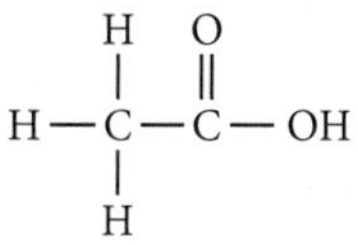 H—C(H)(H)—C(=O)—OH	Ethanoic acid
Ester	O‖ —C—O—R'	alkyl -oate	Esters	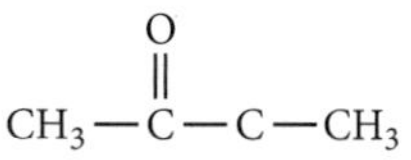 CH_3—C(=O)—C—CH_3	Methyl ethanoate
Amine	$-NH_2$	-anamine	Amines	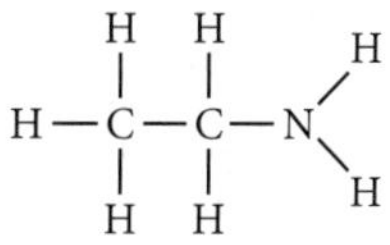 H—C(H)(H)—C(H)(H)—N(H)(H)	Ethanamine
Amide	O‖ —C—NH_2	-anamide	Amides	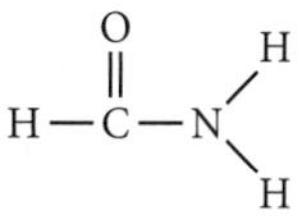 H—C(=O)—N(H)(H)	Methanamide
Nitriles	—C≡N	-nitrile or cyano-	Nitriles/ cyanides	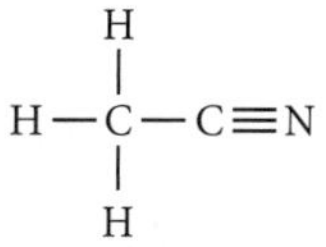 H—C(H)(H)—C≡N	Ethane nitrile

The structures shown in the table are called structural formulas. These show how the atoms in a molecule and the functional group within the molecule are arranged.

Sometimes, when brevity is important, molecules can be represented using condensed formulas known as semi-structural formulas. These do not contain bonds between atoms. Table 4.3 gives some examples of semi-structural formulas.

TABLE 4.3 Structural and semi-structural formulas

Name	Structural formula	Semi-structural formula
Propane	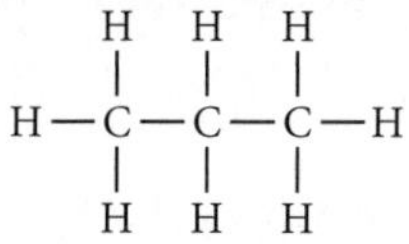 H—C(H)(H)—C(H)(H)—C(H)(H)—H	$CH_3CH_2CH_3$
2-chloro propane	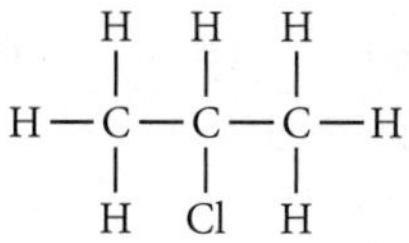 H—C(H)(H)—C(H)(Cl)—C(H)(H)—H	$CH_3CHClCH_3$
Methyl propane	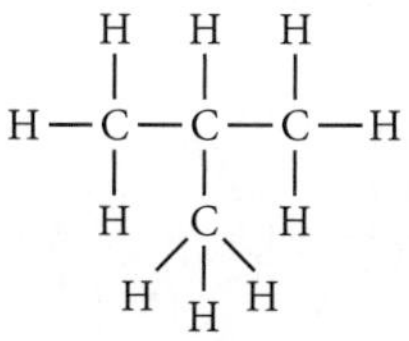 H—C(H)(H)—C(H)(CH_3)—C(H)(H)—H	$CH_3CH(CH_3)CH_3$
Ethanoic acid	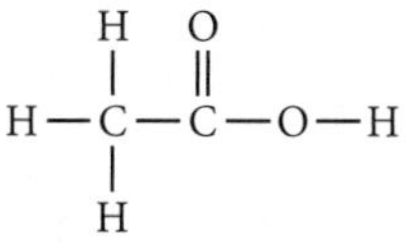 H—C(H)(H)—C(=O)—O—H	CH_3COOH

4.1.2 Naming organic compounds

When naming an organic compound, the following steps are performed.

Alkanes

Alkanes are organic compounds consisting entirely of hydrogen and saturated carbon atoms.

Worked example

Step 1

Determine the number of carbons in the longest carbon chain.

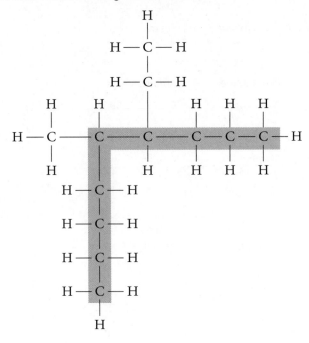

Step 2

Identify any **side groups**.

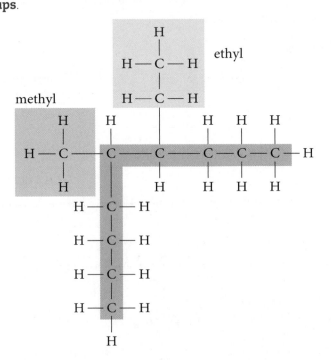

Step 3

Number the carbon chain from the end that gives side groups the smallest numbers.

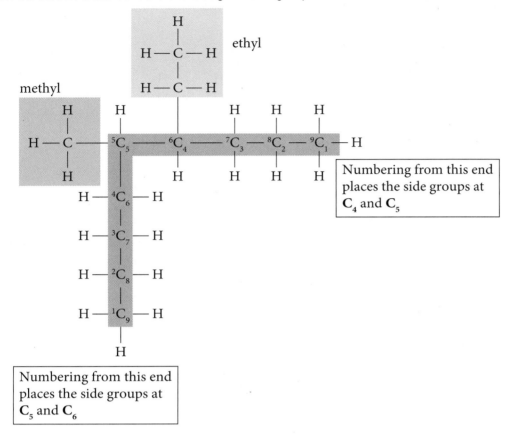

Step 4

Name the compound, remembering to place the side groups in alphabetical order.

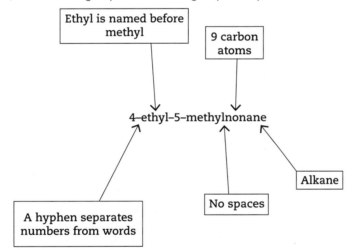

Worked example

Question: Using IUPAC rules, write the name of the compound shown below.

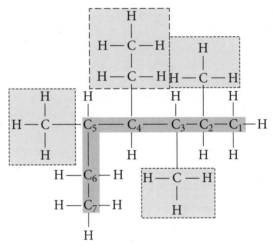

Steps 1–3

Completing steps 1–3 gives:

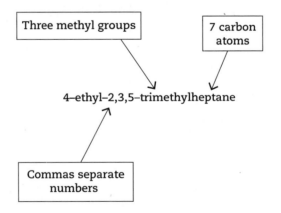

Step 4

Name the compound.

Three methyl groups		7 carbon atoms

4–ethyl–2,3,5–trimethylheptane

Commas separate numbers

Alkenes

An **alkene** contains the carbon=carbon double bond.

$$\begin{array}{c} \diagdown \quad \diagup \\ C = C \\ \diagup \quad \diagdown \end{array}$$

In alkenes, the chain is numbered so the carbon atoms in the double bond have the smallest numbers.

Worked example

Question: Using IUPAC rules, write the name of the compound shown below.

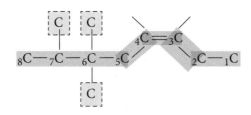

Steps 1–3

Completing steps 1–3 gives the following chain.

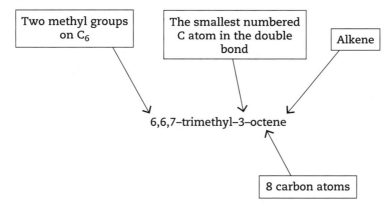

Step 4

Name the compound.

| Two methyl groups on C_6 | The smallest numbered C atom in the double bond | Alkene |

6,6,7–trimethyl–3–octene

8 carbon atoms

Alkynes

Alkynes contain the carbon ≡ carbon triple bond.

$$-C \equiv C-$$

Naming alkynes follows the rules used to name alkenes.

Haloalkanes

Haloalkanes are alkanes that contain one or more halogen atoms (F, Cl, Br, I); for example:

$$
\begin{array}{cc}
\text{H} & \text{Cl} \\
| & | \\
\text{Cl}-\text{C}-\text{H} & \text{Cl}-\text{C}-\text{Cl} \\
| & | \\
\text{H} & \text{H} \\
\text{Chloromethane} & \text{Trichloromethane} \\
& \text{(chloroform)}
\end{array}
$$

When more than one different halogen atoms are present, they are ordered alphabetically; for example:

$$
\begin{array}{ccc}
\text{Cl} & \text{F} & \text{H} \\
| & | & | \\
\text{Cl}-\text{C}- & \text{C}- & \text{C}-\text{H} \\
| & | & | \\
\text{H} & \text{H} & \text{Br}
\end{array}
$$

This compound would be named 1-bromo-3,3-dichloro-2-fluoropropane.

Alcohols

Alcohols are characterised by the −OH group.

A hydrocarbon chain containing an −OH group would be numbered from the end that gave the −OH group the smallest number. This would be named: 3,3-dimethyl-2-butanol.

This can also be named by incorporating the position of the alcohol *within* the alcohol name: 3,3-dimethylbutan-2-ol.

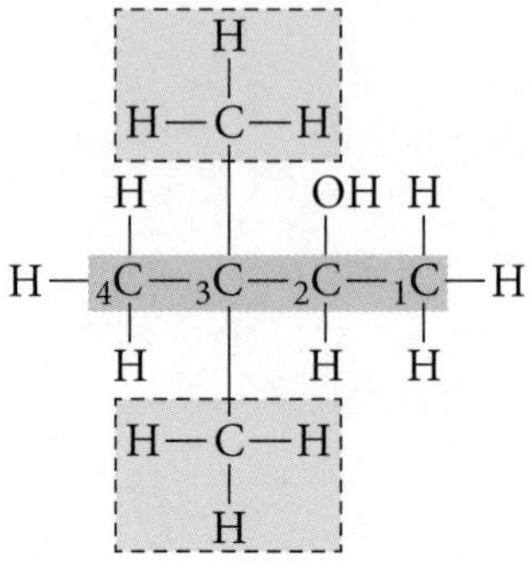

Alcohols can be classed as primary, secondary or tertiary according to how many **alkyl groups** are attached to the carbon atom that is attached to the −OH group.

> **Hint**
>
> **Key concept**
>
> An alkyl group is a carbon side chain such as methyl ($-CH_3$), ethyl ($-CH_2CH_3$) etc. They are represented by the general symbol, R.

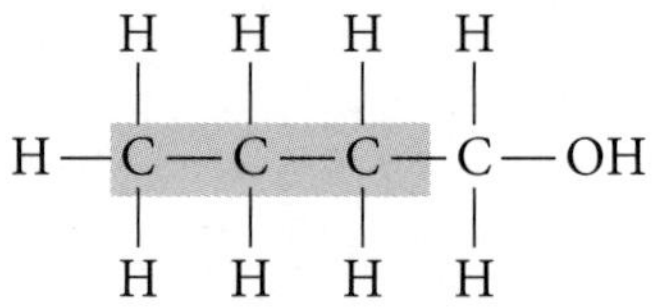

ONE alkyl group: primary
1-butanol

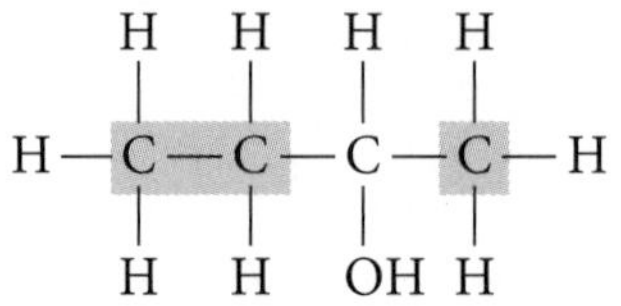

TWO alkyl groups: secondary
2-butanol

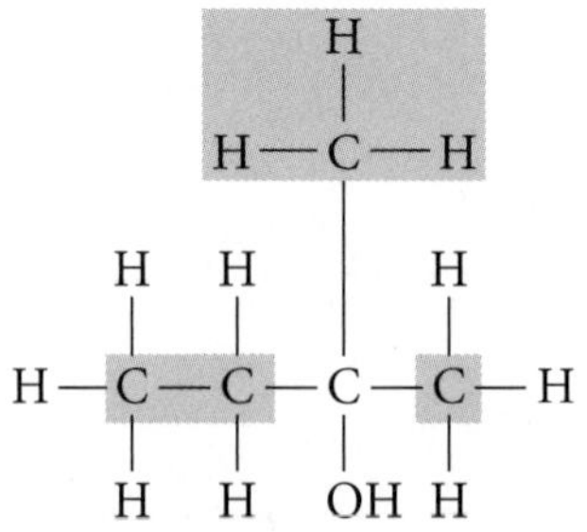

THREE alkyl groups: tertiary
2-methyl-2-butanol

Aldehydes

Aldehydes are a class of compounds containing the carbonyl group, C=O.

$$\begin{array}{c} \diagdown \\ C = O \\ \diagup \end{array}$$

A number of classes of organic compounds contain the carbonyl functional group. The aldehyde class of compounds has the carbonyl group at the *end* of a carbon chain. This is known as a **terminal functional group**.

$$\begin{array}{c} \diagdown \\ C = O \\ H \diagup \end{array}$$

Any aldehyde would be numbered from right to left of the chain, as shown below.

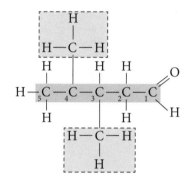

The aldehyde group is always at C_1, so no number is necessary

3,4-dimethylpentanal

al = aldehyde

Ketones

Ketones also contain the carbonyl functional group, but in ketones the carbonyl group is in the *middle* of the carbon chain.

For example:

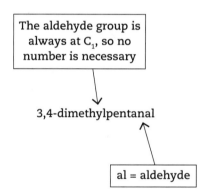

3-methyl-2-hexanone

Carboxylic acids

Carboxylic acids are another class of compounds with a terminal functional group.

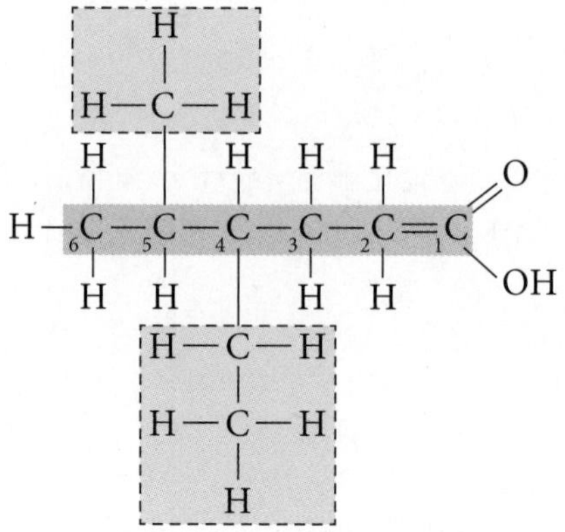

4-ethyl-5-methylhexanoic acid

Esters

Esters are compounds that contain the structural unit:

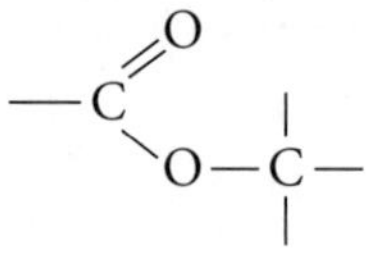

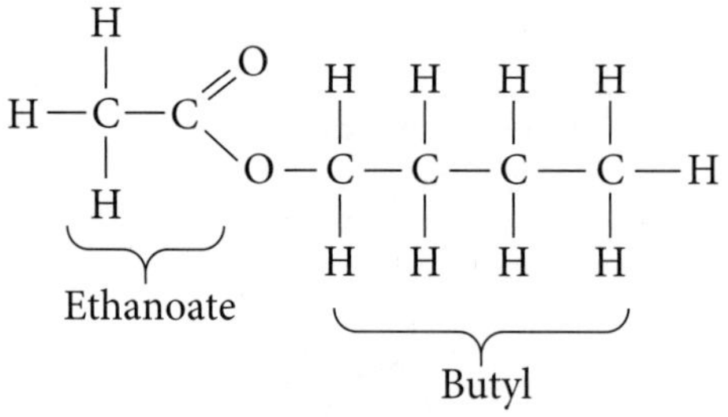

Butyl ethanoate

Nitriles

Nitriles are compounds containing the cyanide functional group $-C{\equiv}N$.

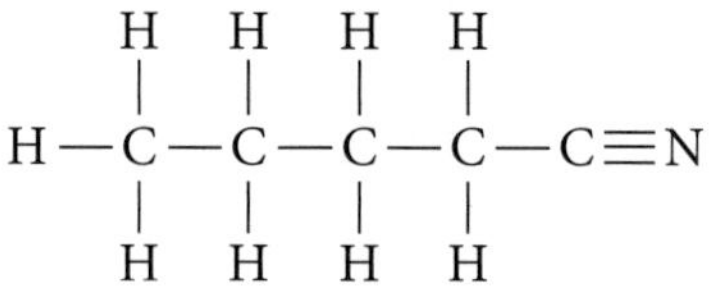

Butanenitrile

Care needs to be taken when the cyanide group is in the middle of a chain. The carbon attached to the N is always C_1.

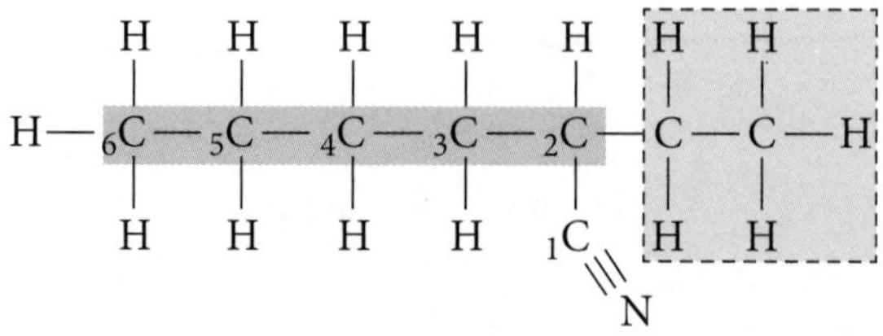

2-ethylhexane nitrile

Amines

A convenient way to think of **amines** is to imagine an ammonia molecule (NH_3) in which one or more of the hydrogen atoms has been replaced by an alkyl group.

Amines can be classed according to how many of the H atoms bonded to the central N atom have been replaced by alkyl groups.

$$H\!\!\diagdown\!\!\underset{H\diagup}{N}\!-\!R \quad \text{Primary, 1°}$$

$$H\!\!\diagdown\!\!\underset{R'\diagup}{N}\!-\!R \quad \text{Secondary, 2°}$$

$$R''\!\!\diagdown\!\!\underset{R'\diagup}{N}\!-\!R \quad \text{Tertiary, 3°}$$

For example, the following molecule would be called butylethylamine. It is a secondary amine.

Amides

Amides contain of a carbonyl group attached to an amine group.

As with amines, amides can be classed according to how many of the H atoms bonded to the central C and N atoms have been replaced by alkyl groups.

When naming amides, it is useful to follow the steps below.

Step 1

Identify the longest chain.

Step 2

The carbonyl carbon is numbered C_1.

For example:

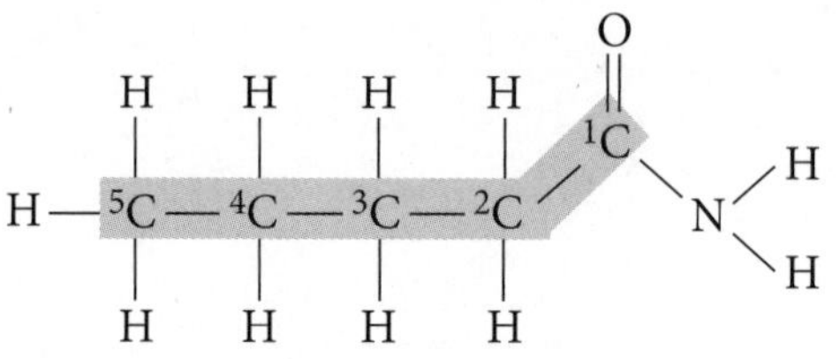

Pentanamide

For more complex amides in which an alkyl group is bonded to the N atom, an extra step is required.

Step 3

N is used to represent an alkyl group bonded directly to the N atom:

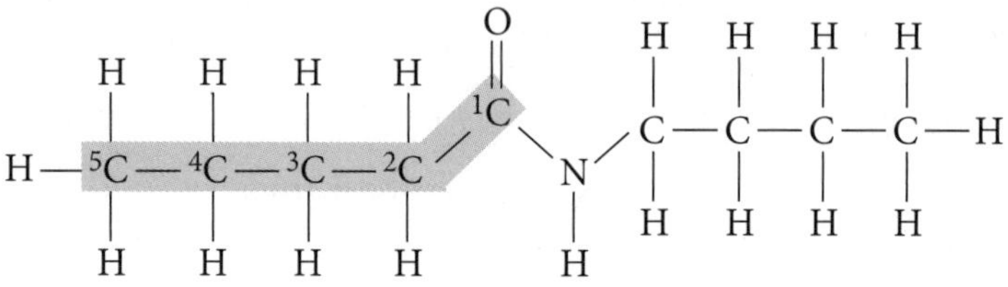

N-butylpentanamide

An example of a tertiary amide is shown below.

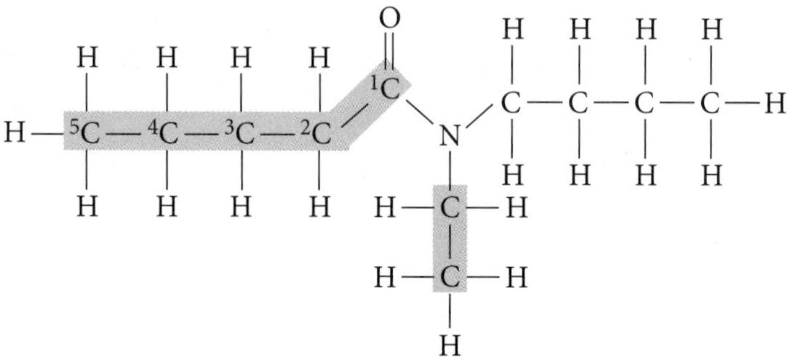

N-butyl-N-ethylpentanamide

4.1.3 Structural isomers

Isomers are molecules with the same **molecular formula** (that is, number of each atom) but different arrangements of atoms. **Structural isomers** are molecules where the number and types of bonds between the atoms, and therefore the arrangement of atoms, is different. For example, the figure that follows shows three structural isomers with the same molecular formula of C_5H_{12}.

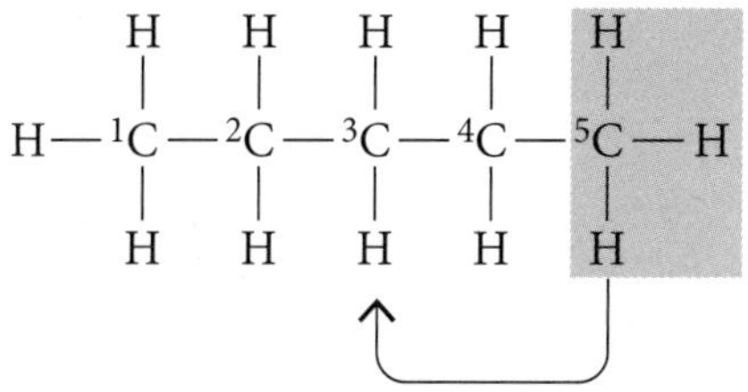

Pentane

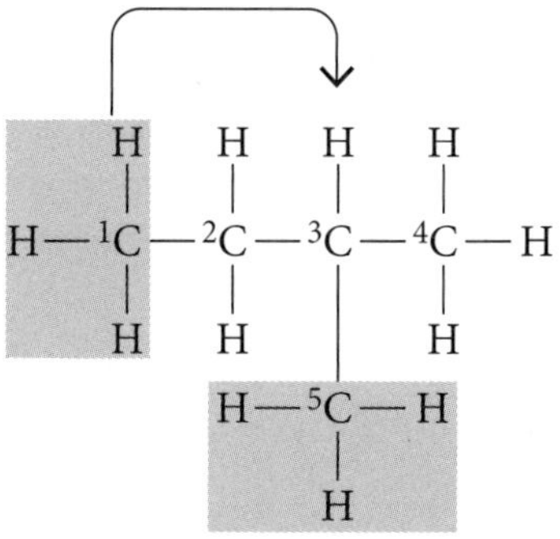

Methylbutane

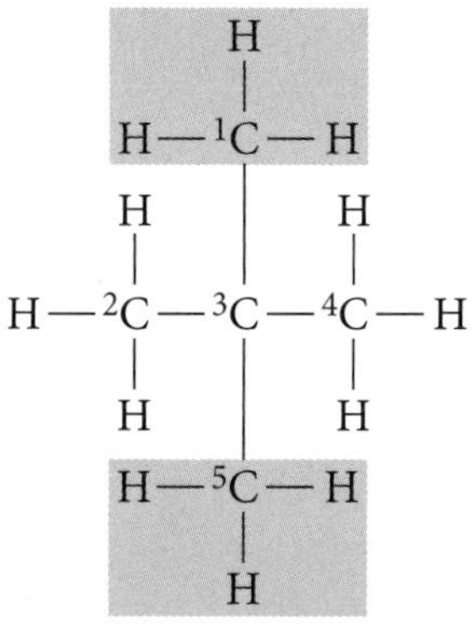

Dimethylpropane

Hint

Hint

Methylbutane COULD have been named 2-methylbutane and dimethylpropane COULD have been named 2,2-dimethylpropane, but the numbers, in these cases, are irrelevant.

4.1.4 Stereoisomers

Stereoisomers are compounds that have the same molecular formula and structural formula but a different arrangement of atoms in three-dimensional space.

A common type of stereoisomerism is **geometrical isomerism**. Geometrical isomers are those that have different arrangements of atoms around a double bond.

Consider the isomers of dichloroethane below.

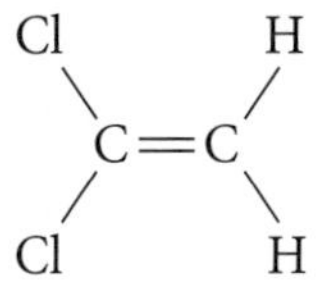

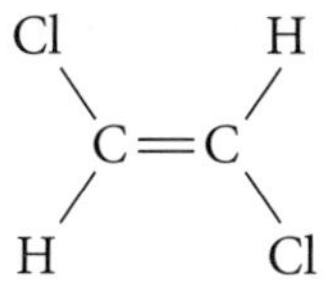

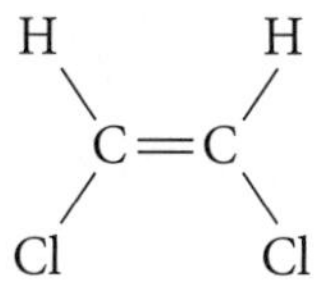

1,1-dichloroethane *trans*-1,2-dichloroethane *cis*-1,2-dichloroethene

While both the second and third isomers have chlorine atoms on different carbon atoms, they are arranged ***differently*** around the double bond.

Hint

When spoken, *cis* sounds like 'sis'. A useful way to remember *cis* and *trans* is: **'Sis'** = **S**ame side of the double bond.

If the H and Cl atoms can be replaced with alkyl groups, this can be applied to long chain alkenes.

Worked example

Question: Write the name of the compound shown below.

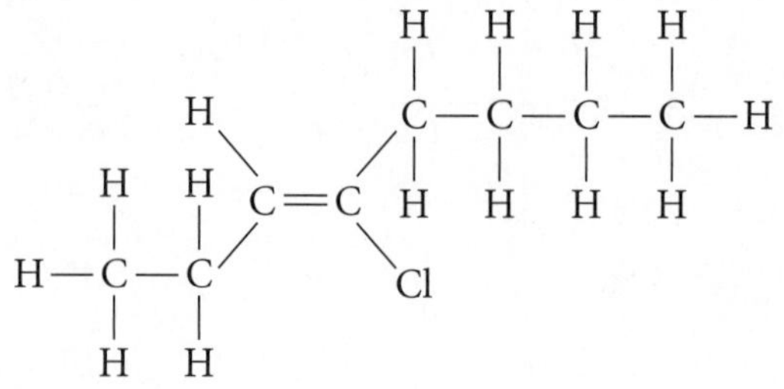

Step 1

Identify the longest carbon chain that contains the double bond. Number the carbon atoms from the side that gives the double bond the smallest number.

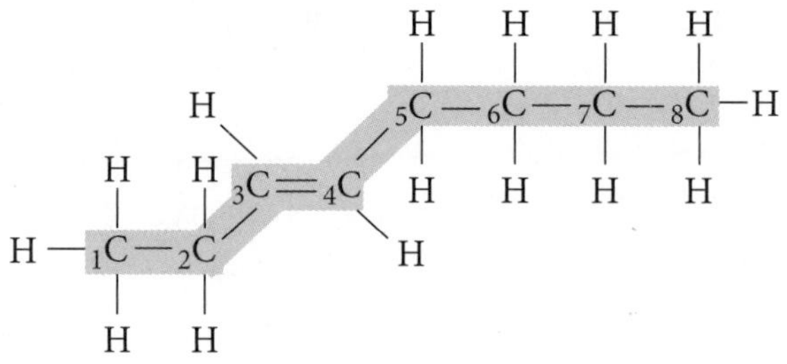

Step 2

Identify the arrangement as *cis* or *trans*. The longest carbon chain goes across the double bond.
This compound would be called *trans*-3-octene.

4.2 Physical properties and trends

4.2.1 Intermolecular forces

The physical properties of organic molecules depend on the nature and strength of the **intermolecular forces** that exist between them.

Dispersion forces

Dispersion forces are the weakest of the intermolecular forces and occur instantaneously between the positive nucleus of an atom in a molecule and the negative electrons of atoms in nearby molecules. Dispersion forces occur between every molecule.

Dipole–dipole interaction

Dipole–dipole interaction occurs in substances that contain polar molecules. These are molecules that contain polar bonds arranged to give a resulting dipole. The shape of molecules is important here. Tetrachloromethane contains the polar C–Cl bond but, because they are arranged in a tetrahedral shape, the polar bonds cancel.

Hydrogen bonding

Hydrogen bonding is the strongest intermolecular force and occurs in substances that contain a hydrogen atom bonded directly to a fluorine atom, an oxygen atom or a nitrogen atom.

> **Hint**
> A good way to recognise hydrogen bonding is: **H–FON**.

4.2.2 Trends in melting and boiling points

Remember, the stronger the intermolecular force in a substance, the more energy is required to separate the molecules of the substance. This means the melting point and boiling point will be higher.

Effect of molecular size and shape

In straight chain alkanes, the larger the molecule, the stronger the dispersion forces, because more atoms are present.

Branched-chain alkanes are more complicated. The diagram below shows the dispersion forces acting between two octane molecules. A star (∗) attached to an H atom shows the potential for forming dispersion forces between other octane molecules.

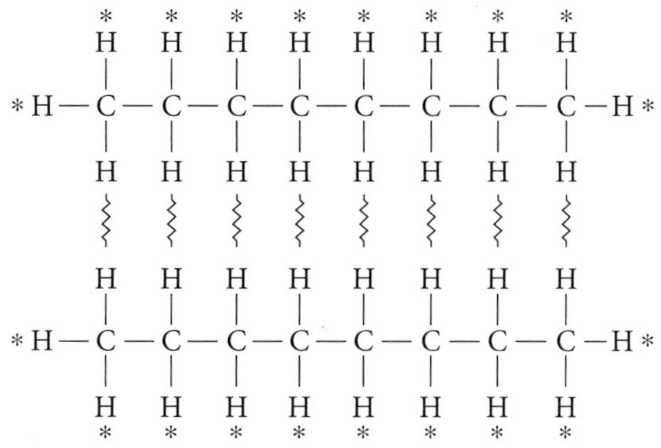

A structural isomer of octene such as 3,4-dimethylhexane is shown below.

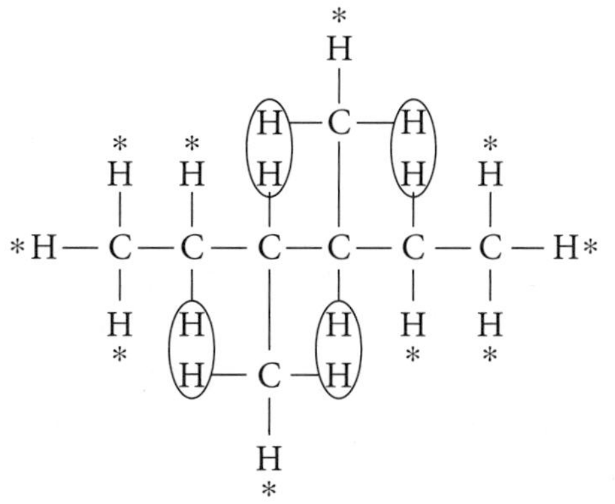

In a branched molecule such as this, the H atoms of the side groups are very close to the H atoms of the main chain. These effectively 'cancel' each other out. This reduces the number of points where dispersion forces can operate. This affects the relative boiling points; for example:

octane = 125.6°C > 3,4-dimethylhexane = 118°C

Effect of functional groups

The type of functional group present in a compound can greatly affect its physical properties. The table below shows the boiling points for molecules in four different homologous series.

TABLE 4.4 Boiling points in four different homologous series

Number of carbon atoms	Boiling points (°C)			
	Alkanes	Alkenes	Alcohols	Carboxylic acids
1	−164		65	100
2	−89	−104	78	118
3	−42	−48	97	141
4	−0.5	−6	118	163
5	36	30	138	187
6	69	64	157	205
7	98	94	178	222
8	125	121	195	240
9	151	146	214	256
10	174	172	229	270

Below is a graph of boiling point trends against the number of carbon atoms, in different homologous series.

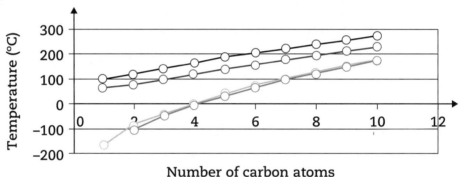

-○- **Boiling points of alkanes**   -○- **Boiling points of alkenes**
-○- **Boiling points of alcohols**   -○- **Boiling points of carboxylic acids**

From the figure above it can be seen that the boiling points of carboxylic acids and alcohols are significantly greater than those of their equivalent alkane and alkene.

Also, as the number of carbon atoms increases in the alkanes and alkenes, their boiling points come closer to those of their alcohol and carboxylic acid equivalents. This is because as the molecules get bigger, the strength dispersion forces increase. The number of polar groups (alcohol and carbonyl) remains the same in larger molecules; therefore, the effect of hydrogen bonding becomes less significant.

4.2.3 Trends in solubility

The intermolecular forces present in organic molecules determines their solubility in water and organic solvents.

Other factors to consider are the size of the molecule, the position of any functional groups and branching.

> **Hint**
> The general rule when discussing solubility is: *Like dissolves like.*

4.2.4 Alcohols

Alcohols contain the hydroxyl functional group, –OH. According to the H–FON rule, alcohols are governed by hydrogen bonding. Water also exhibits hydrogen bonding. Generally, therefore, alcohols are soluble in water.

Consider three alcohols: 1-pentanol, 2-pentanol and 2-methyl-1-butanol. The figure below shows their respective structures and water solubilities.

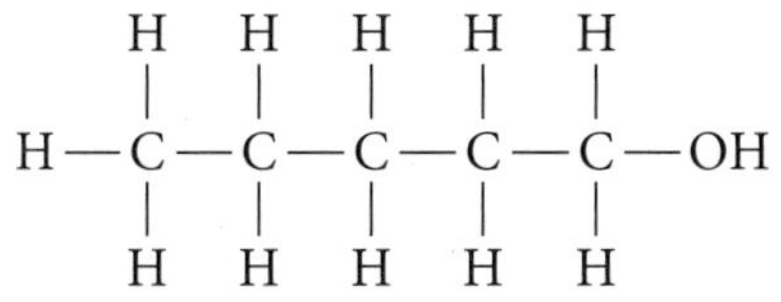

1-pentanol
Solubility = 22 g L^{-1}

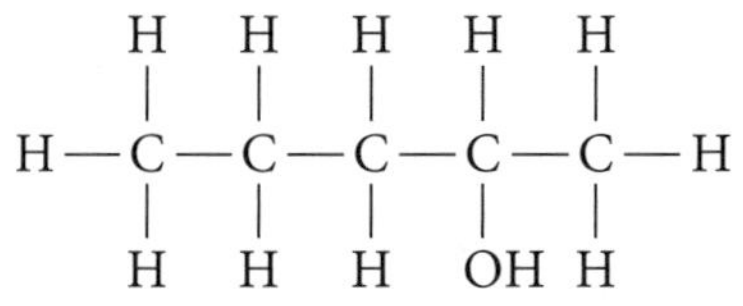

2-pentanol
Solubility = 45 g L^{-1}

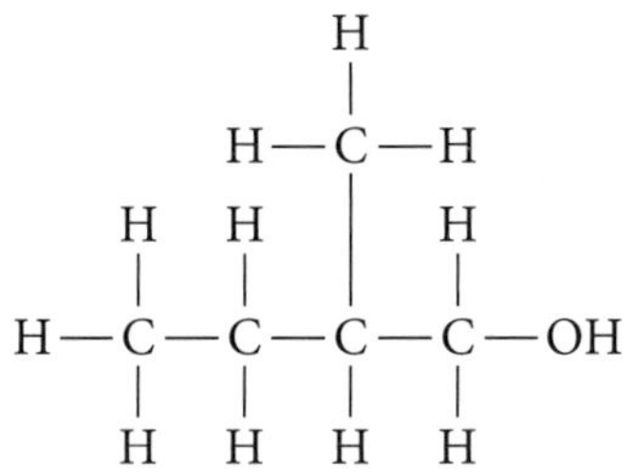

2-methyl-1-butanol
Solubility = 31 g L^{-1}

The –OH group at the end of the chain decreases solubility, whereas branching increases solubility.

4.2.5 Aldehydes and ketones

Aldehydes and ketones both contain the carbonyl functional group. The predominant intermolecular force between these compounds is dipole–dipole interaction. Because of this, aldehydes and ketones have lower boiling points than their alcohol equivalents and are more **volatile**.

4.2.6 Carboxylic acids

These generally have high boiling points because both the carbonyl group and the hydroxyl group of a carboxylic acid participate in hydrogen bonding.

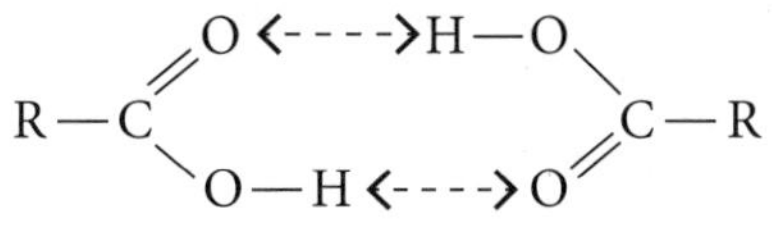

4.2.7 Halides

The predominant intermolecular force present in between haloalkanes is dipole–dipole interaction.

- The larger the halide present, the stronger the intermolecular force.
- The longer the carbon chain, the stronger the intermolecular forces.

4.3   Organic reactions and reaction pathways

The chemical properties of organic compounds depend, for the most part, on the functional group present. In fact, the types of reactions that compounds undergo can be used as a basis for identifying those compounds.

4.3.1 Reactions of alkanes

Combustion

Alkanes, such as methane and octane, shown below, combust to give CO_2 and H_2O.

$$CH_4(g) + 2O_2(g) \rightarrow CO_2(g) + 2H_2O(l)$$

$$2C_8H_{18}(l) + 25O_2(g) \rightarrow 16CO_2(g) + 18H_2O(l)$$

Substitution

Surprisingly, alkanes are chemically quite unreactive. Aside from combustion, the other reaction of note that alkanes undergo is free radical substitution. This involves replacing a hydrogen atom in an alkane with, usually, a halogen.

These reactions require a significant amount of energy in the form of ultraviolet light.

In the reaction below, a chlorine molecule is split by UV light into two chlorine radicals.

| Methane | Chlorine | | Chloromethane | Hydrochloric acid |

One of the chlorine radicals, being very unstable and, hence, reactive, attaches itself to the carbon atom of the methane. One of the C–H bonds of the methane molecule is broken, allowing the Cl radical to take its place in the methane molecule. The resulting H radical forms a bond with the chlorine molecule to give a hydrogen chloride molecule, HCl.

> **Hint**
> Substitution reactions of alkanes are easily recognised because they *always* involve UV light in some way.

4.3.2 Reactions of alkenes

Compounds that contain multiple bonds between carbon atoms (alkenes and alkynes) are referred to as unsaturated. Alkanes only contain single bonds between carbon atoms. These are referred to as saturated.

Alkenes are reactive hydrocarbons due to the presence of their double bond.

Addition reactions

Addition of hydrogen, H_2

Addition reactions involve a molecule being *added* across the double bond.

Consider the following reaction between ethene and hydrogen.

Ethene   Ethane

This reaction, also referred to as **hydrogenation** or reduction, is useful for converting alkanes to alkenes.

Addition of a halogen

This type of reaction is commonly used as a test for unsaturated compounds.

Ethene   1,2-dichloroethane

The test consists of mixing the unknown compound with orange-coloured bromine water. If the unknown compound is unsaturated, the orange mixture is decolourised.

Addition of water (hydration)

This reaction converts an alkene into an alcohol.

Ethene   Ethanol

Addition of hydrogen halide, HX

Hydrogen halides are HF, HCl, HBr and HI.

Ethene   Bromoethane

Worked example

Question: Draw and name the product formed when 1-butene reacts with hydrogen chloride.

2-chlorobutane   1-chlorobutane

When this reaction is carried out, both products would be formed but there would be a greater proportion of 2-chlorobutane.

This can be predicted using Markovnikov's rule.

Hint

Key concept
Markovnikov's rule:
In an addition reaction of hydrogen halides or water to unsymmetrical alkenes, the hydrogen atom is added preferentially to the carbon atom from the C=C bond that is already bonded to the greater number of hydrogen atoms.

Polymerisation

Due to the presence of the double bond, alkenes are able to react with themselves to form long chains.

This type of reaction is called **addition polymerisation** and is discussed in more detail on page 125.

4.4 Reactions of haloalkanes

The C−X bond of a haloalkane is polar, with the halogen side of the bond being slightly more negative than the carbon side. Other species that are negatively charged, such as the hydroxide ion, OH^-, or possess a lone pair of electrons, such as ammonia, NH_3, can form bonds with the slightly positive carbon atom in preference to the halogen. These substances that can replace the halogen atom are called nucleophiles and the reaction is called **nucleophilic substitution**.

Reaction with sodium hydroxide

$$H-\overset{\overset{\displaystyle H}{|}}{\underset{\underset{\displaystyle H}{|}}{C}}-\overset{\overset{\displaystyle H}{|}}{\underset{\underset{\displaystyle H}{|}}{C}}-\overset{\overset{\displaystyle H}{|}}{\underset{\underset{\displaystyle H}{|}}{C}}-Br \; + \; OH^- \; \rightarrow \; H-\overset{\overset{\displaystyle H}{|}}{\underset{\underset{\displaystyle H}{|}}{C}}-\overset{\overset{\displaystyle H}{|}}{\underset{\underset{\displaystyle H}{|}}{C}}-\overset{\overset{\displaystyle H}{|}}{\underset{\underset{\displaystyle H}{|}}{C}}-OH \; + \; Br^-$$

1-bromopropane   1-propanol

Reaction with ammonia

$$H-\overset{\overset{\displaystyle H}{|}}{\underset{\underset{\displaystyle H}{|}}{C}}-\overset{\overset{\displaystyle H}{|}}{\underset{\underset{\displaystyle H}{|}}{C}}-\overset{\overset{\displaystyle H}{|}}{\underset{\underset{\displaystyle H}{|}}{C}}-Br \; + \; NH_3 \; \rightarrow \; H-\overset{\overset{\displaystyle H}{|}}{\underset{\underset{\displaystyle H}{|}}{C}}-\overset{\overset{\displaystyle H}{|}}{\underset{\underset{\displaystyle H}{|}}{C}}-\overset{\overset{\displaystyle H}{|}}{\underset{\underset{\displaystyle H}{|}}{C}}-NH_2 \; + \; HBr$$

1-bromopropane   1-propanamine

Reaction with cyanide (as KCN solution)

$$H-\overset{\overset{\displaystyle H}{|}}{\underset{\underset{\displaystyle H}{|}}{C}}-\overset{\overset{\displaystyle H}{|}}{\underset{\underset{\displaystyle H}{|}}{C}}-\overset{\overset{\displaystyle H}{|}}{\underset{\underset{\displaystyle H}{|}}{C}}-Br \; + \; CN^- \; \rightarrow \; H-\overset{\overset{\displaystyle H}{|}}{\underset{\underset{\displaystyle H}{|}}{C}}-\overset{\overset{\displaystyle H}{|}}{\underset{\underset{\displaystyle H}{|}}{C}}-\overset{\overset{\displaystyle H}{|}}{\underset{\underset{\displaystyle H}{|}}{C}}-C\equiv N \; + \; Br^-$$

1-bromopropane   1-butanenitrile

An important reaction of nitriles is their hydrogenation (reduction) to form amines.

1-butanenitrile → 1-butanamine (H$_2$/Pd)

Elimination reactions of haloalkanes

These types of reactions convert the haloalkane back to an alkene (with the elimination of the HX molecule).

1-bromopropane → Propene + HBr (NaOH/Ethanol)

4.4.1 Reactions of alcohols

Combustion

Alcohols burn readily in air to produce carbon dioxide and water.

$$CH_3CH_2OH(l) + 3O_2(g) \rightarrow 2CO_2(g) + 3H_2O(g)$$

Dehydration (elimination)

Alcohols also undergo **dehydration** by losing water to form alkenes when heated with concentrated sulfuric or phosphoric acid catalysts. This is a form of elimination, where one larger molecule is converted into two smaller molecules; for example:

$$CH_3CH_2CHOHCH_3(l) \rightarrow CH_3CH_2CH=CH_2(g) + H_2O(g)$$
2-butanol · 1-butene

Oxidation

This is a very important class of reactions that can be used to identify alcohols.

Alcohols undergo **oxidation** with strong oxidising agents such as acidified potassium permanganate or potassium dichromate solutions.

> **Hint**
>
> Acidified potassium permanganate:
>
> $$MnO_4^-(aq) + 8H^+(aq) + 5e^- \rightarrow Mn^{2+}(aq) + 4H_2O(l)$$
>
> Acidified potassium dichromate:
>
> $$Cr_2O_7^{2-}(aq) + 14H^+(aq) + 6e^- \rightarrow 2Cr^{3+}(aq) + 7H_2O(l)$$
> orange · green
>
> These half-equations for the reduction of permanganate and dichromate in acid conditions do NOT have to be memorised. They are in the table of standard electrode potentials on page 10 of your *Chemistry formula and data booklet*.

The products of the oxidation of alcohols depend on the type of alcohol: primary, secondary or tertiary.

Primary alcohols undergo oxidation in two stages to give an aldehyde followed by a carboxylic acid.

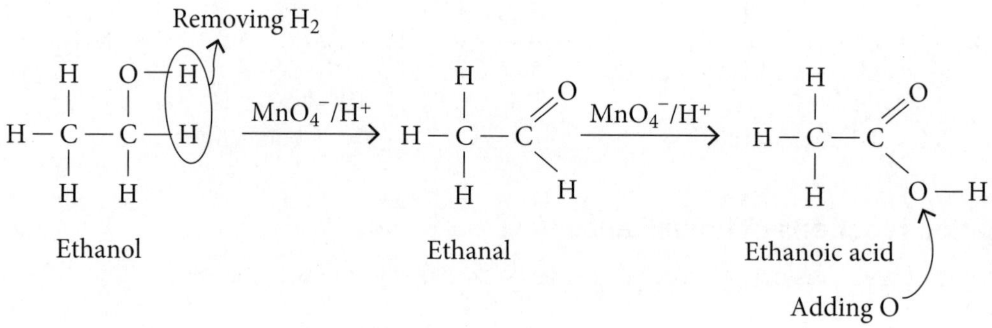

Secondary alcohols undergo oxidation in one stage to give a ketone.

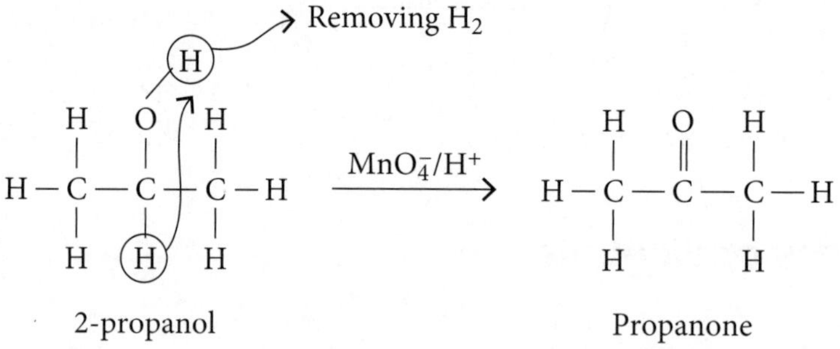

Tertiary alcohols are not oxidised by permanganate of dichromate solutions.

> **Hint**
>
> Primary and secondary alcohols decolourise acidified MnO_4^- and acidified $Cr_2O_7^{2-}$ solutions. Tertiary alcohols have no effect.

Reaction of alcohols with sodium

This is a similar type of reaction to that between an acid and an active metal.

$$alcohol + metal \rightarrow salt + hydrogen$$
$$2CH_3CH_2\text{–}O\text{–}H + Na \rightarrow 2CH_3CH_2\text{–}O^-Na^+ + H\text{–}H$$
$$\text{sodium ethoxide}$$

Production of esters (esterification)

Alcohols react with carboxylic acids to form esters and water.

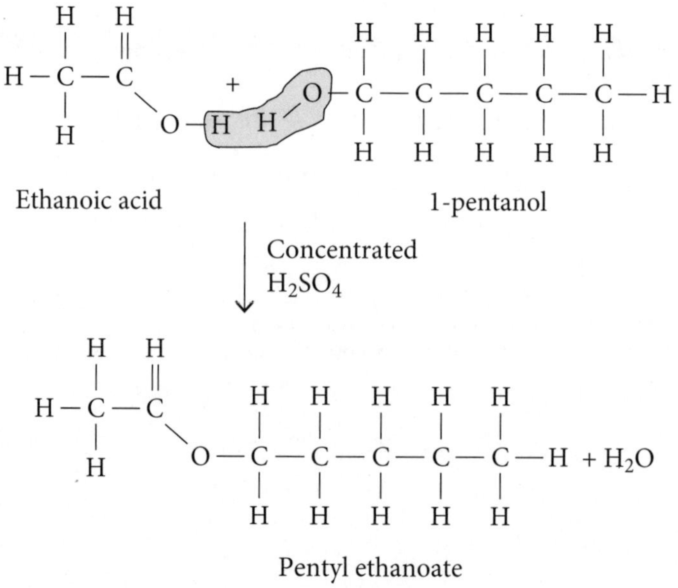

This reaction produces a water molecule. Any reaction, including the one above, in which the functional groups of two molecules join and a small water molecule is released is called a **condensation reaction**.

Esters are an important group of organic compounds found in a wide range of biological systems. They have pleasant, fruity odours. The ester shown here, pentyl ethanoate, has a smell reminiscent of bananas.

The reaction shown above is reversible. Under certain conditions, esters can react with water to form the original acid and alcohol from which they were made.

4.4.2 Reactions of carboxylic acids

Carboxylic acids undergo reactions that are normally associated with acids.

- Reactions with reactive metals, such as magnesium:

$$Mg(s) + 2CH_3COOH(aq) \rightarrow (CH_3COO)_2Mg(aq) + H_2(g)$$

- Reactions with bases are neutralisation reactions, creating a salt and water:

$$NaOH(aq) + CH_3COOH(aq) \rightarrow CH_3COONa(aq) + H_2O(l)$$

- Reactions with carbonates:

$$CaCO_3(s) + 2CH_3COOH(aq) \rightarrow (CH_3COO)_2Ca(aq) + CO_2(g) + H_2O(l)$$

This is a useful reaction to test for the presence of carboxylic acids by adding a carbonate and noting the evolution of bubbles of carbon dioxide gas.

4.4.3 Reactions of amines and amides

As with alcohols and haloalkanes, amines may be classified as primary, secondary or tertiary, depending on the number of alkyl groups attached to the nitrogen. A primary amine has one alkyl group, a secondary amine has two alkyl groups attached to the nitrogen, and a tertiary amine has three alkyl groups attached to the nitrogen. The figure below shows examples of these.

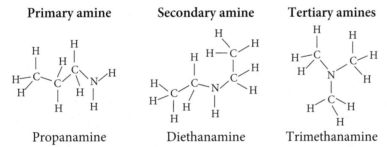

Primary amine	Secondary amine	Tertiary amines
Propanamine	Diethanamine	Trimethanamine

Reaction with inorganic acids

Amines are weak bases and, as such, will undergo a neutralisation reaction with acids to form a salt and water.

$$R-NH_2(aq) + H^+(aq) \rightarrow R-NH_3^+(aq)$$

For example:

$$CH_3NH_2(aq) + HCl(aq) \rightarrow CH_3NH_3^+Cl^-(aq)$$

Methanamine   Methylammonium chloride

Condensation reaction with carboxylic acids

$$
\underset{\text{Ethanoic acid}}{
\begin{array}{c}
\text{H} \quad \text{O} \\
| \qquad \| \\
\text{H}-\text{C}-\text{C} \\
| \qquad \diagdown \\
\text{H} \qquad \boxed{\text{O}-\text{H}}
\end{array}}
\quad + \quad
\underset{\text{1-pentanamine}}{
\begin{array}{c}
\quad\; \text{H} \;\; \text{H} \;\; \text{H} \;\; \text{H} \;\; \text{H} \\
\quad\quad | \quad | \quad | \quad | \quad | \\
\textcircled{H}\diagdown\text{N}-\text{C}-\text{C}-\text{C}-\text{C}-\text{C}-\text{H} \\
\quad\; | \quad\; | \quad | \quad | \quad | \quad | \\
\quad\; \text{H} \;\; \text{H} \;\; \text{H} \;\; \text{H} \;\; \text{H}
\end{array}}
$$

$$
\underset{\text{N-pentylethanamide}}{
\begin{array}{c}
\text{H} \quad \text{O} \\
| \qquad \| \qquad\quad \text{H} \;\; \text{H} \;\; \text{H} \;\; \text{H} \;\; \text{H}\\
\text{H}-\text{C}-\text{C} \qquad\; | \quad | \quad | \quad | \quad | \\
| \qquad \diagdown \text{N}-\text{C}-\text{C}-\text{C}-\text{C}-\text{C}-\text{H} + \text{H}_2\text{O}\\
\text{H} \qquad | \quad\; | \quad | \quad | \quad | \quad | \\
\quad\; \text{H} \;\; \text{H} \;\; \text{H} \;\; \text{H} \;\; \text{H} \;\; \text{H}
\end{array}}
$$

4.4.4 Reaction pathways

Sometimes the production of a finished product requires a series of reactions from a particular reactant. This is known as a **reaction pathway**.

Worked example

Question: Describe a reaction pathway for the formation of N-pentylbutanamide from 1-bromobutane.

Step 1

Draw the structures of the reactant and product.

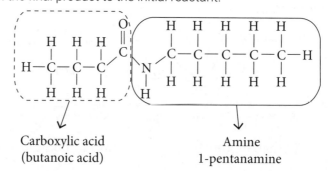

1-bromobutane   N-pentylbutanamide

Step 2

Work backwards from the final product to the initial reactant.

Carboxylic acid
(butanoic acid)   Amine
                           1-pentanamine

Step 3

Track each of these final reactants separately.

 Butanoic acid is produced by the oxidation of a primary alcohol.

1-butanol $\xrightarrow{MnO_4^-/H^+}$ Butanoic acid

1-Butanol is produced from the reaction between 1-bromobutane.

1-bromobutane $\xrightarrow{NaOH}$ 1-butanol   + NaBr

1-Pentanamine is produced from the reduction (hydrogenation) of a nitrile.

1-pentanenitrile $\xrightarrow{H_2/Pd}$ 1-pentanamine

1-Pentanenitrile is produced from the reaction between 1-bromobutane and potassium cyanide.

1-bromobutane $\xrightarrow{KCN}$ 1-pentanenitrile

4.5   Organic materials: structure and function

4.5.1 Polymers

Polymers are giant molecules, often called macromolecules, which form when many thousands of smaller molecules, called **monomers**, are joined by covalent bonds. The process of linking monomers together is called **polymerisation**. If the polymer consists of more than one different monomer, it is known as a **copolymer**.

Structure of polymers

The structure of polymers can be described in four ways that reflect increasing complexity.

Name	Structural formula
Primary structure This is the sequence of monomers in the polymer chain.	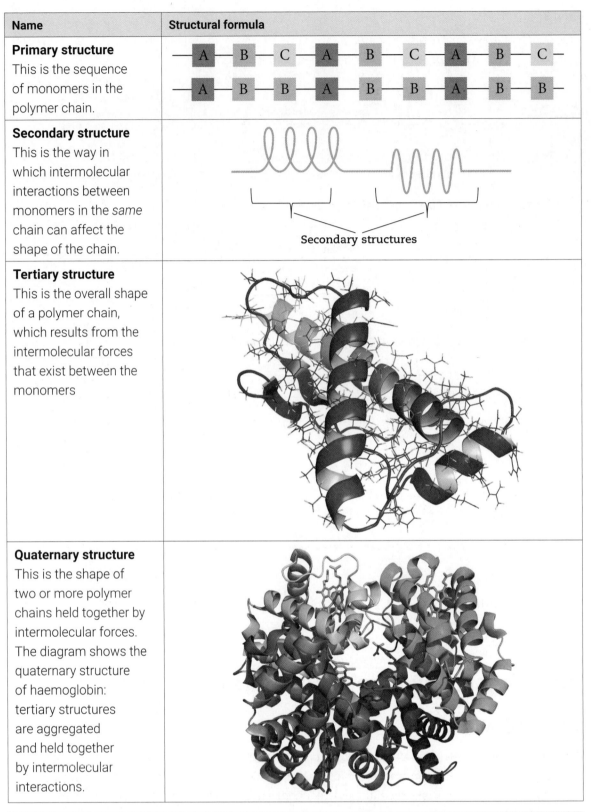
Secondary structure This is the way in which intermolecular interactions between monomers in the *same* chain can affect the shape of the chain.	
Tertiary structure This is the overall shape of a polymer chain, which results from the intermolecular forces that exist between the monomers	
Quaternary structure This is the shape of two or more polymer chains held together by intermolecular forces. The diagram shows the quaternary structure of haemoglobin: tertiary structures are aggregated and held together by intermolecular interactions.	

FIGURE 4.1 The structure of polymers

The properties of polymers

The variety in the shape and size of polymers affects their properties. Physical properties include biodegradability, tensile strength, and density.

Biodegradability

Biodegradability is the ability of a polymer to decay naturally as a result of enzymes produced by bacteria.

Generally, natural polymers are more likely to be biodegradable.
Synthetic polymers, such as polyethene and polypropene, are not
biodegradable.

Tensile strength

This is the resistance of a polymer to being pulled or stretched. In certain
applications, this is a desirable property; when manufacturing nylon ropes,
for example. Tensile strength can be achieved in a number of ways, such as
introducing 'cross-linking' between chains to keep them in place.

FIGURE 4.2 Cross-linking of polymers

Density

The density of a polymer depends on how tightly packed the polymer chains are. This can be affected
by the presence of any side chains in the polymer and if the chain exhibits branching.

4.5.2 Proteins

Proteins are large polymers (**polyamides** or **polypeptides**) formed when many smaller monomer
molecules link together in a condensation polymerisation reaction. The monomer building blocks of
proteins are **amino acids**. Although there are only 20 different amino acids found in proteins, there
are hundreds of thousands of different proteins.

It is important to understand how amino acids link together to
form the primary structure, how the polymer chain is arranged to give
its secondary structure and how the polymer chains interact to
give the proteins its tertiary and quaternary structures.

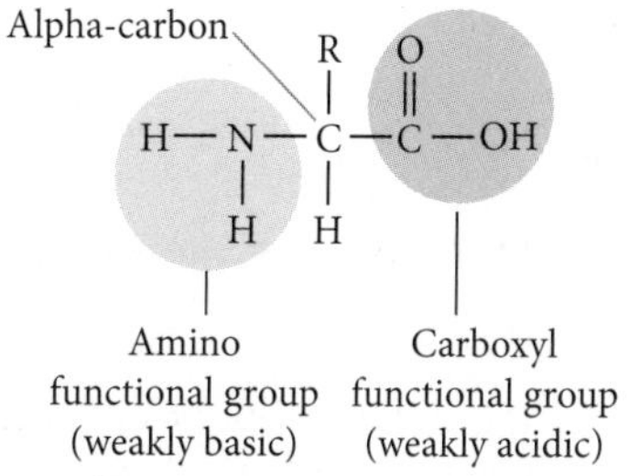

Primary structure

The monomers that link together to form a chain are called
amino acids. Figure 4.3 shows the general structure of an
alpha-amino acid.

FIGURE 4.3 Structure of an alpha-amino acid

> **Hint**
>
> The 20 amino acids commonly found in proteins are listed on pages 12 and 13 of the *Chemistry formula and data booklet*.
> Some of the structures may seem intimidating but they *all* have the structure shown above. The only difference is the R group.

Zwitterions

Due to the presence of a −COOH group (which tends to lose its H to form the −COO⁻ ion) and a −NH$_2$
group (which tends to gain an H to form the −NH$_3^+$ ion), amino acids exist in a **zwitterion** form.

$$H_2N-CH-COOH \rightleftharpoons {}^+H_3N-CH-COO^-$$

In acidic conditions, the zwitterion accepts a H⁺:

$$^+H_3N-CH-COO^- + H^+ \longrightarrow {}^+H_3N-CH-COOH$$

In basic conditions, the zwitterion loses a H⁺:

$$^+H_3N-CH-COO^- + OH^- \longrightarrow H_2N-CH-COO^- + H_2O$$

The pH at which an amino acid exists as an overall neutral zwitterion is called the **isoelectric point**.
This value is unique to each amino acid and appears next to each amino acid in the *Chemistry formula
and data booklet*. Isoelectric points become very important in the analysis of proteins.

Secondary structure

This arises from bonding between amino acids within a chain and with amino acids in nearby chains.

There are two main arrangements:

The −C=O group of one amino acid forms a hydrogen bond to an −NH group *four* peptide links further along the chain.

This produces a spiral structure called an α-**helix**.

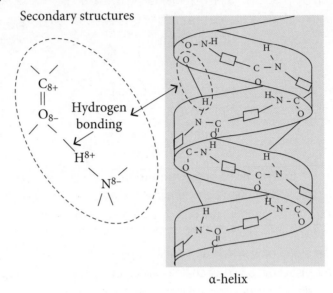

FIGURE 4.4 α-helix and secondary structures

When stretched, regions of different chains lie alongside each other, and hydrogen bonding may occur between chains, resulting in a β-**pleated sheet** structure.

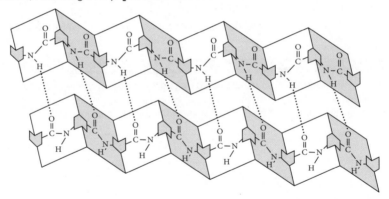

FIGURE 4.5 β-pleated sheet

9780170459150

Tertiary structure

The tertiary structure of a protein is the three-dimensional shape that forms due to interactions between the secondary structures present. Various types of interactions that occur between the R group side chains on the amino acids that make up the chain are summarised below.

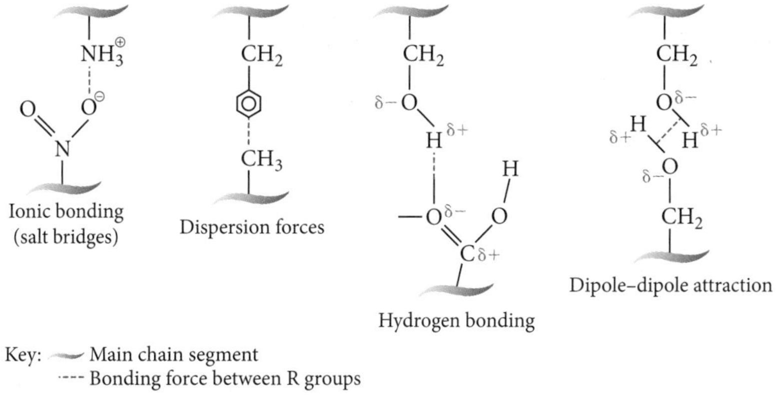

Ionic bonding
(salt bridges)

Dispersion forces

Hydrogen bonding

Dipole–dipole attraction

Key: — Main chain segment
    ---- Bonding force between R groups
Note: —OH groups can also interact by hydrogen bonding

FIGURE 4.6 Three-dimensional interactions between secondary structures of proteins

Quarternary structure

If there is more than one polypeptide chain involved, then there may be a quaternary structure, where the same types of intermolecular attractions cause the polypeptide chains to aggregate together in a very specific conformation. It is this three-dimensional quaternary structure of the overall protein that allows it to interact with other molecules. For example, an enzyme must have a three-dimensional structure that allows it to interact specifically with its substrate; and haemoglobin (made of four polypeptide subunits) must have the correct three-dimensional structure so it can interact with iron and oxygen.

4.5.3 Enzymes

Enzymes are usually large protein molecules. They are critical for speeding up many of the body's chemical reactions. An example is the breaking down of a sucrose molecule into glucose and fructose:

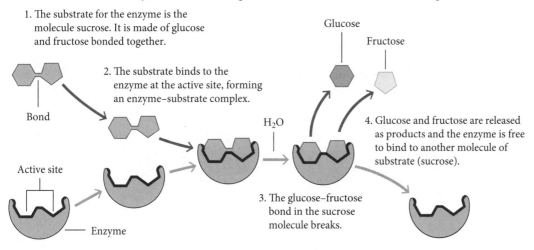

1. The substrate for the enzyme is the molecule sucrose. It is made of glucose and fructose bonded together.

2. The substrate binds to the enzyme at the active site, forming an enzyme–substrate complex.

Bond

Active site

Enzyme

H_2O

3. The glucose–fructose bond in the sucrose molecule breaks.

Glucose

Fructose

4. Glucose and fructose are released as products and the enzyme is free to bind to another molecule of substrate (sucrose).

FIGURE 4.7 Enzymes breaking down sucrose molecule into glucose and fructose

4.5.4 Carbohydrates

Carbohydrates are natural polymers that are made up of three elements – carbon, hydrogen and oxygen.

The most common formula for carbohydrates is $C_6H_{12}O_6$.

Complex carbohydrates such as starch and cellulose are made up of simple units called **monosaccharides**.

Monosaccharides

These are the simple sugars and are **all** isomers of $C_6H_{12}O_6$. They have the empirical formula CH_2O. Figure 4.8 shows structures of three common monosaccharides. The left-hand structures show all the atoms of the molecules; in the right-hand ones, the carbon atoms are not labelled.

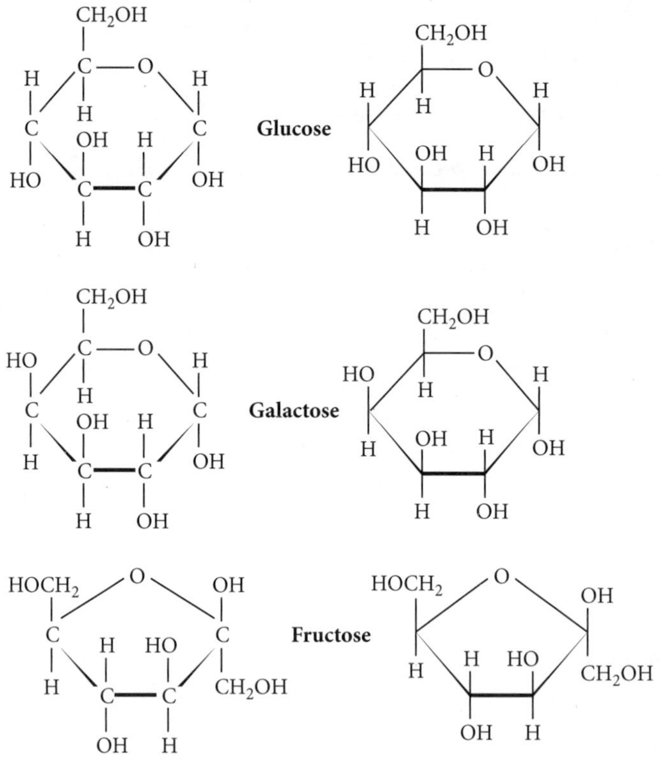

FIGURE 4.8 The monosaccharides glucose, galactose and fructose

> **Hint**
>
> The structures for glucose and fructose appear alongside their straight-chain forms on page 11 of your *Chemistry formula and data booklet*.

Figure 4.9 shows these monosaccharides in their straight-chain forms. In this form it can be seen that glucose and galactose possess the aldehyde group, –CHO. Consequently, glucose and galactose are referred to as aldose sugars. Fructose contains the ketone group, –C=O, and is known as a ketose sugar.

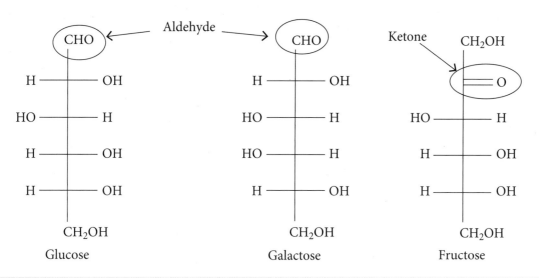

FIGURE 4.9 The monosaccharides glucose, galactose and fructose in their straight-chain forms

There are two isomers of glucose, called α-glucose and β-glucose, shown in Figure 4.10. The α form of glucose has the −OH group on C_1, below the plane of the ring, whereas in β-glucose it is above the plane of the ring.

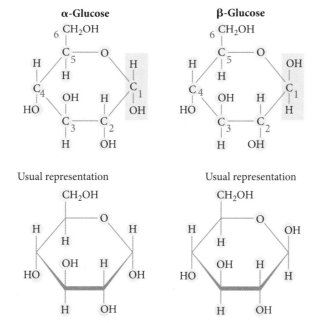

FIGURE 4.10 α-glucose and β-glucose

Disaccharides

The reaction that occurs to form a **disaccharide** is a condensation reaction, because a water molecule is released. This can be written simply as:

$$2C_6H_{12}O_6 \rightarrow C_{12}H_{22}O_{11} + H_2O$$

Consider the formation of the disaccharide sucrose (sugar) from α-glucose and β-fructose shown in Figure 4.11.

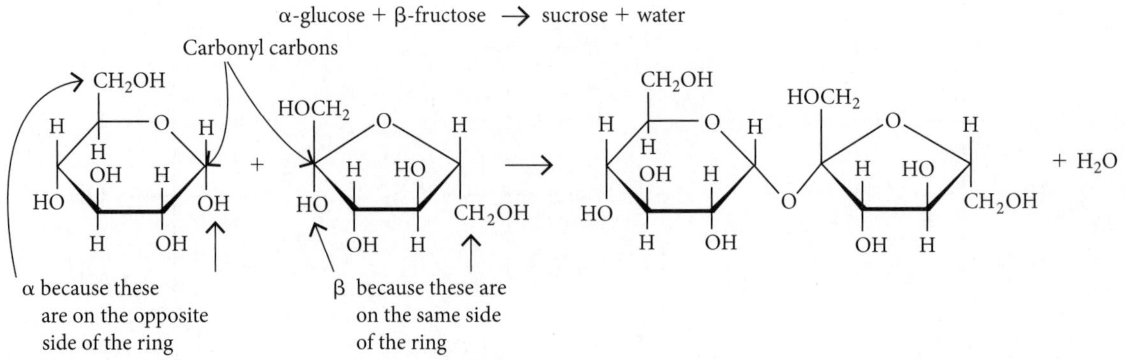

FIGURE 4.11 The formation of disaccharide sucrose (sugar) from α-glucose and β-fructose

Polysaccharides

Polysaccharides are formed when hundreds or thousands of monosaccharides join through condensation polymerisation.

The three polysaccharides important to human life are starch, cellulose and glycogen.

Starch

Starch is an important nutrient because it provides a source of energy for the biological processes that occur in the human body.

Starch consists of long chains of α-glucose monomers. These chains take two forms.

Amylose is a straight chain arrangement formed from α-glycosidic linkages between the C_1 atom of one glucose molecule and C_4 atom of an adjacent glucose molecule.

Amylopectin is a branched-chain arrangement formed from α-glycosidic linkages between the C_1 atom of one glucose molecule and C_6 atom of an adjacent glucose molecule in addition to the 1 to 4 linkages found in amylose.

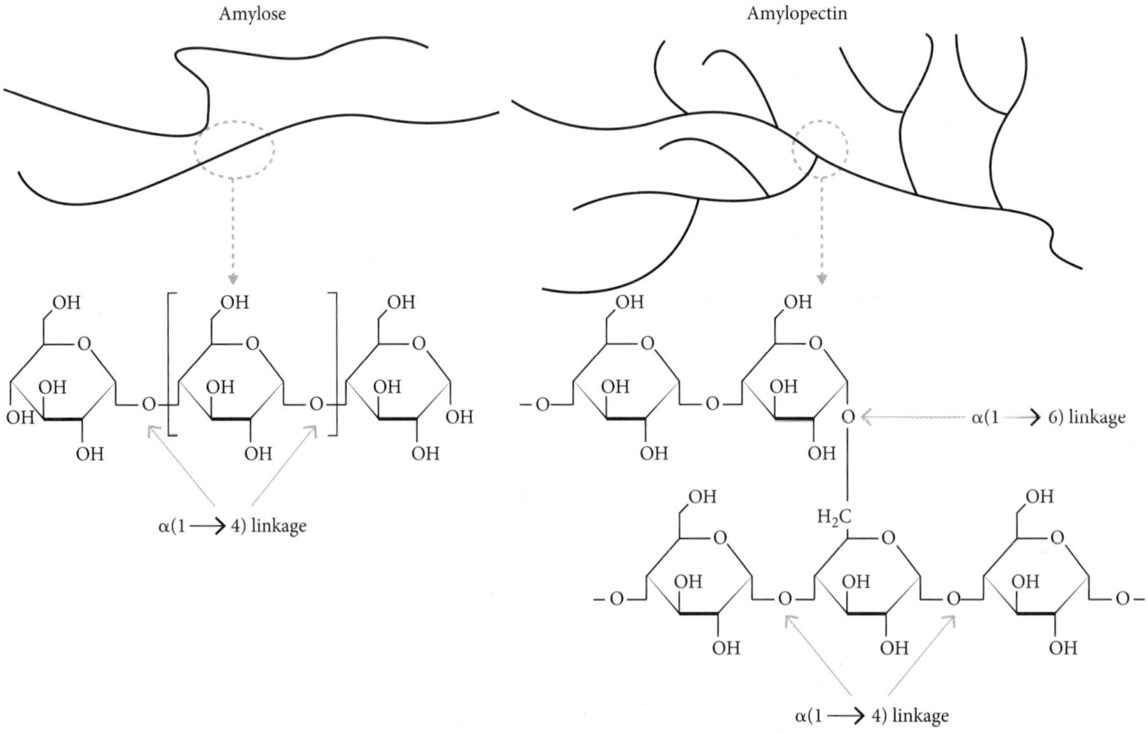

FIGURE 4.12 Amylose and amylopectin

Cellulose

Cellulose is formed from β-glucose molecules joined through 1 to 4 glycosidic linkages. Cellulose is an important part of a balanced diet for humans as it is the main component of dietary fibre.

The β-glucose molecules are joined in straight chains and oriented such that strong hydrogen bonds are formed between chains, to form a highly cross-linked structure.

Glycogen

Glycogen is a polymer of glucose and is used by animals to store the energy released by respiration.

Its structure is similar to amylopectin but with a slightly different frequency of branching.

4.5.5 Lipids

Lipids are naturally occurring ester molecules and belong to the organic family called **triglycerides**. If they are solid at room temperature, then they are called fats. If they are liquid at room temperature, then they are oils.

Lipids are formed when 1,2,3-propanetriol (glycerol) reacts with fatty acid molecules.

> **Hint**
>
> A fatty acid is a carboxylic acid with a long hydrocarbon R group, which is CH_{12} in this example:
>
> $CH_3(CH_2)_{12}COOH$

Production of a fat

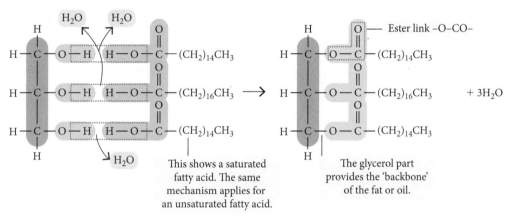

Glycerol + 3 fatty acids $\xrightarrow[\text{reaction}]{\text{Condensation}}$ Triglyceride + $3H_2O$

Hydrolysis of a fat

Triglyceride + $3H_2O$ $\xrightarrow[\text{reaction}]{\text{Hydrolysis}}$ Glycerol + 3 fatty acids

FIGURE 4.13 The production and hydrolysis of a lipid

Fatty acids can be saturated, all C–C single bonds, or unsaturated, one or more C=C double bonds.

Saponification

Saponification is the production of soaps through the hydrolysis of fats.

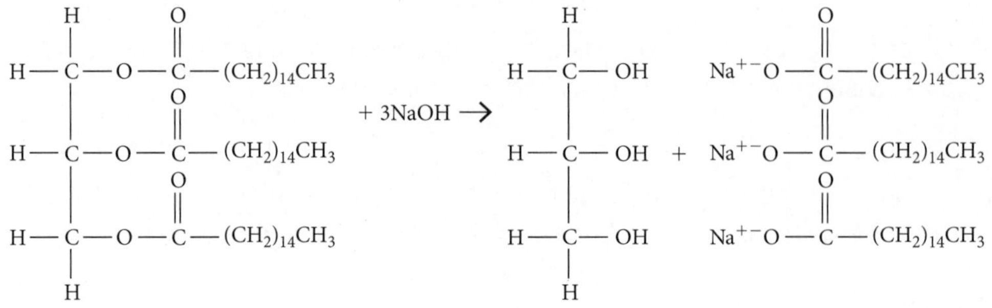

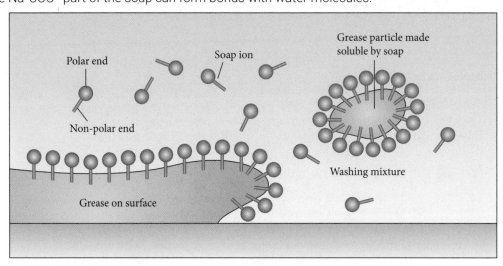

FIGURE 4.14 Saponification of a lipid

Soaps are able to clean because they can bond with polar and non-polar substances. The long hydrocarbon chains are **hydrophobic** but they can form bonds with non-polar dirt and grease. The Na^+COO^- part of the soap can form bonds with water molecules.

FIGURE 4.15 The cleaning action of soap

4.5.6 The structure and properties of synthetic polymers

It is important to distinguish between synthetic polymers and plastic. While all plastics are synthetic polymers, not all synthetic polymers are plastics. Plastics are malleable and pliable and capable of being moulded by heat and pressure.

LDPE and HDPE

These are different forms of polyethene, one of the most important synthetic polymers. LDPE is **L**ow **D**ensity **P**oly**E**thene and HDPE is **H**igh **D**ensity **P**oly**E**thene.

HDPE contains a large number of polymer chains in a given volume. This is because they are straight-chain polymers which allow them to be packed close together. This produces a polymer with high density, high melting point and low flexibility.

LDPE has relatively few polymer chains in a given volume. This is because it possesses significant chain branching. This produces a polymer with low density, low melting point and greater flexibility.

Polypropene

Polypropene is an addition polymer much like polyethene, but with a $-CH_3$ side group.

Whereas polymerisation of ethene produces a straight chain polyethene molecule, polymerisation of propene produces a straight chain polymer but with side groups positioned along the chain.

$$
\begin{array}{ccccc}
\text{H} \ \text{H} & \text{H} \ \text{H} & \text{H} \ \text{H} & \text{H} \ \text{H} & \text{H} \ \text{H} \\
| \ | & | \ | & | \ | & | \ | & | \ | \\
\text{C}=\text{C} + & \text{C}=\text{C} + & \text{C}=\text{C} + & \text{C}=\text{C} + & \text{C}=\text{C} \\
| \ | & | \ | & | \ | & | \ | & | \ | \\
\text{H} \ \text{H} & \text{H} \ \text{H} & \text{H} \ \text{H} & \text{H} \ \text{H} & \text{H} \ \text{H}
\end{array}
\qquad
\begin{array}{ccccc}
\text{H} \ \text{H} & \text{H} \ \text{H} & \text{H} \ \text{H} & \text{H} \ \text{H} & \text{H} \ \text{H} \\
| \ | & | \ | & | \ | & | \ | & | \ | \\
\text{C}=\text{C} + & \text{C}=\text{C} + & \text{C}=\text{C} + & \text{C}=\text{C} + & \text{C}=\text{C} \\
| \ | & | \ | & | \ | & | \ | & | \ | \\
\text{H} \ \text{CH}_3 & \text{H} \ \text{CH}_3 & \text{H} \ \text{CH}_3 & \text{H} \ \text{CH}_3 & \text{H} \ \text{CH}_3
\end{array}
$$

$$\downarrow \qquad\qquad\qquad\qquad \downarrow$$

Polyethene   Polypropene

FIGURE 4.16 Polymerisation of propene

The arrangement of the $-CH_3$ groups in polypropene is very important. There are three variations in the structure of polypropene depending on the orientation of the $-CH_3$ group:

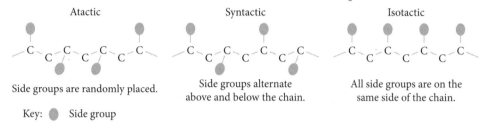

Atactic   Syntactic   Isotactic

Side groups are randomly placed.   Side groups alternate above and below the chain.   All side groups are on the same side of the chain.

Key: ● Side group

FIGURE 4.17 Three possible arrangements of side groups in polypropene

The *atactic* arrangement results in weaker dispersion forces between chains. This results in a soft polymer with a low melting point.

The *syntactic* arrangement enables close packing of the chains, resulting in stronger intermolecular forces.

The *isotactic* arrangement enables the closest packing of the chains. This results in the strongest of the intermolecular forces.

Polytetrafluoroethene (PTFE)

PTFE is formed in a similar way to polyethene, via addition polymerisation.

$$
n \left(\begin{array}{cc} \text{F} & \text{F} \\ | & | \\ \text{C}=\text{C} \\ | & | \\ \text{F} & \text{F} \end{array} \right) \longrightarrow
\begin{array}{ccccc}
\text{F} & \text{F} & \text{F} & \text{F} & \text{F} \\
| & | & | & | & | \\
-\text{C}-\text{C}-\text{C}-\text{C}-\text{C}- \\
| & | & | & | & | \\
\text{F} & \text{F} & \text{F} & \text{F} & \text{F}
\end{array}
$$

FIGURE 4.18 Formation of polytetrafluoroethene (PTFE)

This polymer has strong intermolecular forces between the chains, which make it strong and tough.

4.6 Analytical techniques

4.6.1 Chromatography

Chromatography is useful for analysing large organic molecules such as proteins. Chromatographic techniques include column, ion exchange and size exclusion chromatography.

Column chromatography

This uses a **stationary phase** and a **mobile phase**. The amount of time a sample to be analysed spends in the column is dependent on the sample's affinity for the mobile phase.

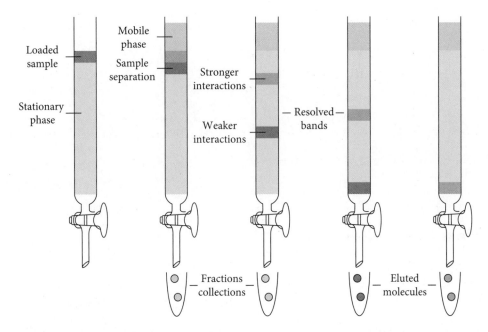

FIGURE 4.19 Column chromatography

Other similar techniques have been developed specifically for protein analysis.

Ion exchange chromatography

As mentioned earlier in this chapter, a unique property that proteins possess is their isoelectric point. This can be used to separate and analyse proteins.

> **Hint**
>
> The isoelectric point (pI) of a protein is the pH at which there is no net charge on a protein.

In ion exchange chromatography, the protein mixture is placed into a **buffer solution**. A buffer solution maintains a constant pH when small amounts of acid and base are added.

So, if a mixture of two proteins with pI values of 4.8 and 6.1 are to be separated, a buffer solution with a pH value somewhere in between would be necessary – 5.2 pH would be suitable.

When protein 1 (pI = 4.8) is mixed with the buffer solution, it will behave like an acid because its pI is lower than the pH of the buffer. The result will be that protein 1 will be negatively charged.

When protein 2 (pI = 6.1) is mixed with the buffer solution, it will behave like a base because its pI is higher than the pH of the buffer. The result will be that protein 2 will be positively charged.

When the protein mixture is introduced to a column containing a cation exchange resin, the *positively* charged protein 2 molecules (cations) will stick to the resin, allowing protein 1 molecules to pass straight through, thereby separating them.

Size exclusion chromatography

This is a filtering technique that depends on the relative abilities of molecules to pass through pores in the stationary phase.

4.6.2 Electrophoresis

In **gel electrophoresis**, separation of large organic molecules such as proteins is based on the mobility of ions in an electric field and so is dependent on their net charge, size and shape.

So, if a mixture of proteins is placed into wells at one end of a gel that contains a buffer solution, the proteins will become charged in a similar way to that discussed in ion exchange chromatography.

Once the proteins are ionised, they interact with the electric field set up within the gel and they begin to move through the gel. The size, shape and amount of charge on the protein molecules affect how fast they move through the gel.

The buffer solutions at either end of the gel have different pH values. This sets up a pH gradient across the gel.

Eventually a protein molecule will move to an area of the gel that has the same pH as its pI. At this point the protein loses its charge and becomes neutral. Because it is neutral, it is no longer soluble and precipitates out into a band that can be detected, often using a UV light.

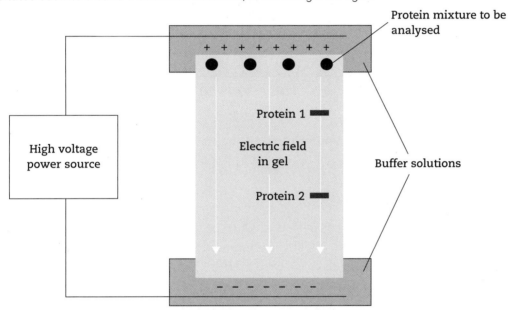

FIGURE 4.20 Schematic diagram of electrophoresis

4.6.3 Mass spectrometry

In addition to providing information about the molar mass and the molecular formula of a compound, mass spectrometry can give information about the structure of a molecule.

In mass spectrometry, a vaporised sample is injected into a section in which the sample is ionised by removing some of its electrons with an electron beam.

The positively charged sample particle is then accelerated by an electric field along a curved tube. Further along this tube, the charged particle encounters a powerful magnetic field which causes it to deflect from its original path. The amount of deflection is dependent on a particle's charge and its mass. The particle then moves into a detector where it is recorded.

When high energy electrons are fired at an organic compound, they can cause the molecule to fragment. These fragments can lead to a distinct fragmentation pattern in the mass spectrum.

Organic molecule fragmentation patterns are useful in determining the structure of the molecule.

When ethanol is injected into the mass spectrometer a mass spectrum is produced (see Figure 4.21 below).

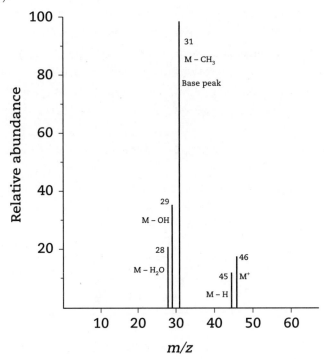

FIGURE 4.21 Mass spectrum of ethanol

The molecular mass of ethanol is given by the small peak at 46. This corresponds to the molecular ion, M^+.

The largest peak is called the base peak and is assigned a relative abundance of 100%.

The peak at 45 corresponds to a loss of H which often occurs when an −OH group is present.

Another common fragment is the $-CH_3$. If this is removed from an ethanol molecule, the result is the $-CH_2OH$ fragment that corresponds to the base peak at 31.

The peak at 29 corresponds to the loss of −OH.

The peak at 28 corresponds to the loss of H_2O. Again, this is a common fragment when an −OH group is present.

Mass spectrometry is a powerful tool for the elucidation of the structure of organic compounds but is best used as a technique to complement other methods.

4.6.4 Infrared spectroscopy

This is an analytical method that uses photons of infrared light to give information about the bonding in organic molecules.

Atoms in molecules are always moving and it is useful to think of chemical bonds as springs which are constantly bending, stretching and twisting. When photons of infrared light interact with these chemical bonds, the photons lose a little of their energy. This energy loss is different for the type of bond and the atoms in the bond. These energy losses can be measured and presented as an infrared or IR spectrum.

Figure 4.22 shows the IR spectra of a number of compounds and their 'signature' peaks – those that can be used to identify the functional groups present.

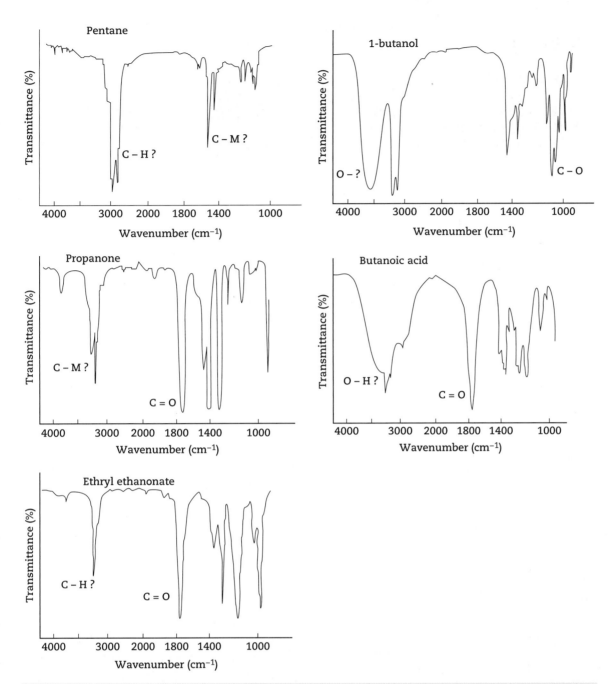

FIGURE 4.22 IR spectra of pentane, 1-butanol, propanone, butanoic acid and ethryl ethanonate and their 'signature' peaks

4.6.5 X-ray crystallography

X-ray crystallography provides information about the relative positions of atoms in a molecule. For this to occur, the substance must be in solid, crystalline form. It has proven to be an extremely accurate method for determining the structures of proteins, enzymes, drugs and viruses.

When X-rays are fired at a crystal, the atoms present scatter the X-rays because the distance between the atoms is roughly the same as the wavelength of the X-rays. The scattered X-rays are detected and produce a **diffraction pattern**. The diffraction pattern is analysed by computers to give very precise information about the structure of the substance.

Glossary

α–helix (or alpha helix)
a structure of natural proteins where the carbonyl (C=O) of one amino acid is hydrogen bonded to the amino H (N—H) of an amino acid that is four down the chain, pulling it into a helical structure, like a curled ribbon

β-pleated sheet (or beta-pleated sheet)
two or more segments of a polypeptide chain lie alongside one another, to form a sheet-like structure held together by hydrogen bonds

addition polymerisation
formation of polymer chains by the addition reaction of unsaturated monomers

alcohol
an organic compound containing the hydroxyl functional group (–OH). Can be classed as primary, secondary or tertiary according to how many alkyl groups are attached to the carbon atom, which is attached to the –OH group. Generally soluble in water

aldehyde
an organic compound consisting of the carbonyl functional group, C=O, bonded to a hydrogen on one end of the molecule

alkane
an organic compound consisting of hydrogen and saturated carbon atoms

alkene
an organic compound that contains a carbon=carbon double bond

alkyl group
a carbon side chain such as methyl (–CH$_3$), ethyl (–CH$_2$CH$_3$) etc. It is represented by the general symbol R

alkyne
an organic compound containing a carbon≡carbon triple bond

amide
an organic compound containing a functional group consisting of a carbonyl group attached to an amine group

amine
an ammonia molecule (NH$_3$) in which one or more of the hydrogen atoms has been replaced by an alkyl group

amino acid
an organic compound containing a carboxyl and an amino group

buffer solution
a solution that maintains a constant pH when small amounts of acid and base are added

carbohydrates
a naturally occurring compound with the general formula C$_x$(H$_2$O)$_y$. The most common formula for carbohydrates is C$_6$H$_{12}$O$_6$.

carbon skeleton
the arrangement of carbon atoms that other atoms join to in an organic molecule

carboxylic acid
an acid containing the carboxyl group C(=O)OH attached to an R group

catenate
the ability of carbon atoms to bond with other carbon atoms to form chains and rings

chromatography
a technique for separating mixtures due to differing interactions with different media. Includes column, ion exchange and size exclusion chromatography

condensation reaction
where two molecules combine to form a larger molecule and produce water as a by-product

copolymer
a polymer made from more than one type of monomer

dehydration
an elimination reaction where a molecule of water is lost from an alcohol to form an alkene

diffraction pattern
a pattern that forms as the result of waves being spread when passed through a crystal

dipole–dipole interaction
intermolecular forces that exist between molecules that have a permanent dipole (equal and opposite charge)

disaccharide
a sugar made from two monosaccharide units

dispersion forces
the weakest intermolecular forces that exist between all molecules

functional group
a group of atoms in a molecule that causes the molecule to chemically react in a distinctive way

gel electrophoresis
the process of separating large, charged molecules by placing them in an electric field and observing their subsequent migration through a medium such as a gel

geometrical isomerism
the quality in an isomer of having different arrangements of atoms around a double bond

haloalkane
an alkane that contain one or more halogen atoms (F, Cl, Br, I)

homologous series
a series of organic molecules with the same functional group but different length carbon chains

hydrogenation
an addition reaction where hydrogen is the molecule added

hydrogen bonding
the strong intermolecular forces that exist between molecules that possess O—H, N—H or F—H bonds

hydrophobic
a molecule that does not bond with water

intermolecular forces
forces that act between molecules; these forces largely determine the physical properties of molecules

isoelectric point
the pH value at which there is no net charge on a protein

isomers
molecules with the same molecular formula (that is, number of each atom) but different arrangements of atoms

IUPAC
International Union of Pure and Applied Chemistry. Formed in 1919 to foster international cooperation and international standardisation in chemistry

lipids
naturally occurring ester molecules belonging to the organic family called triglycerides; called fats if solid at room temperature, oils if liquid at room temperature

ketones
a sugar in which a carbonyl group is bonded to two carbon atoms

mobile phase
a liquid or gas that carries the sample through a chromatography column

molecular formula
the number of atoms of each element present in a molecule

monomers
an individual unit of a polymer

monosaccharides
a sugar that cannot be decomposed into simpler sugars

nomenclature
the system for naming chemical substances

nucleophilic substitution
a reaction of haloalkanes where the halogen atom is replaced by an electron-rich species called a nucleophile

oxidation (of alcohols)
removing hydrogen (H_2) and adding oxygen (O)

polyamides
polymers held together by amide bonds – formed between the amino group of one monomer and a carboxyl group of the next monomer

polymers
large molecules made from many thousands of repeating units

polymerisation
the process by which monomers link together to form a polymer

polypeptides
polyamide chains which constitute proteins

polysaccharides
carbohydrate molecules made from many monosaccharide units

proteins
organic compounds made from long chains of amino acids

reaction pathway
a series of reactions leading to a specific outcome or product

stationary phase
a solid to which the sample molecules adsorb as they move through a chromatography column

stereoisomers
two molecules with the same structural formula but a different arrangement of atoms in space

structural isomers
two molecules with the same molecular formula but a different arrangement of atoms

terminal functional group
a functional group that occurs only at the end of a carbon chain, such as an aldehyde or a carboxylic acid

triglyceride
an ester consisting of glycerol and fatty acid molecules

volatile
something that evaporates easily at room temperature

X-ray crystallography
a technique for determining the relative positions of atoms in a molecule by firing X-rays at a substance in crystalline form

zwitterion
the form of an amino acid where the amino group is protonated at the same time as the carboxyl group is deprotonated, so that there is no net charge

Revision summary

Use the following summary of syllabus dot points and key knowledge within Unit 4 Topic 1 to ensure that you have thoroughly reviewed the content. Provide a brief definition or comment for each item to demonstrate your understanding or code them using the traffic light system – Green (all good); Amber (needs some review); Red (priority area to review).

Structure of organic compounds	
• recognise that organic molecules have a hydrocarbon skeleton and can contain functional groups, including alkenes, alcohols, aldehydes, ketones, carboxylic acids, haloalkanes, esters, nitriles, amines, amides, and that structural formulas (condensed and extended) can be used to show the arrangement of atoms and bonding in organic molecules	
• deduce the structural formulas and apply IUPAC rules in the nomenclature of organic compounds (parent chain up to 10 carbon atoms) with simple branching for alkanes, alkenes, alkynes, alcohols, aldehydes, ketones, carboxylic acids, haloalkanes, esters, nitriles, amines and amides	
• identify structural isomers as compounds with the same molecular formula but different arrangement of atoms; deduce the structural formulas and apply IUPAC rules in the nomenclature for isomers of the non-cyclic alkanes up to C_6	
• identify stereoisomers as compounds with the same structural formula but with different arrangement of atoms in space; describe and explain geometrical (*cis* and *trans*) isomerism in non-cyclic alkenes.	
• Mandatory practical: Construct 3D models of organic molecules.	››

››	**Physical properties and trends**	
	• recognise that organic compounds display characteristic physical properties, including melting point, boiling point and solubility in water and organic solvents that can be explained in terms of intermolecular forces (dispersion forces, dipole–dipole interactions and hydrogen bonds), which are influenced by the nature of the functional groups	
	• predict and explain the trends in melting and boiling points for members of a homologous series	
	• discuss the volatility and solubility in water of alcohols, aldehydes, ketones, carboxylic acids and halides.	
Organic reactions and reaction pathways		
	• appreciate that each class of organic compound displays characteristic chemical properties and undergoes specific reactions based on the functional group present; these reactions, including acid–base and oxidation reactions, can be used to identify the class of the organic compound	
	• understand that saturated compounds contain single bonds only and undergo substitution reactions, and that unsaturated compounds contain double or triple bonds and undergo addition reactions	
	• determine the primary, secondary and tertiary carbon atoms in halogenoalkanes and alcohols and apply IUPAC rules of nomenclature	

››

» • describe, using equations:
  – oxidation reactions of alcohols and the complete combustion of alkanes and alcohols
  – substitution reactions of alkanes with halogens
  – substitution reactions of haloalkanes with halogens, sodium hydroxide, ammonia and potassium cyanide
  – addition reactions of alkenes with water, halogens and hydrogen halides
  – addition reactions of alkenes to form poly(alkenes)	

• recall the acid–base properties of carboxylic acids and explain, using equations, that esterification is a reversible reaction between an alcohol and a carboxylic acid	
• recognise the acid–base properties of amines and explain, using equations, the reaction with carboxylic acids to form amides	
• recognise reduction reactions and explain, using equations, the reaction of nitriles to form amines and alkenes to form alkanes	

• recognise and explain, using equations, that:
  – esters and amides are formed by condensation reactions
  – elimination reactions can produce unsaturated molecule and explain, using equations, the reaction of haloalkanes to form alkenes	

• understand that organic reactions can be identified using characteristic observations and recall tests to distinguish between: – alkanes and alkenes using bromine water – primary, secondary and tertiary alcohols using acidified potassium dichromate(VI) and potassium manganate(VII)	
• understand that the synthesis of organic compounds often involves constructing reaction pathways that may include more than one chemical reaction	
• deduce reaction pathways, including reagents, conditions and chemical equations, given the starting materials and the product.	

Organic materials: structure and function

• appreciate that organic materials including proteins, carbohydrates, lipids and synthetic polymers display properties including strength, density and biodegradability that can be explained by considering the primary, secondary and tertiary structures of the materials	
• describe and explain the primary, secondary (α-helix and β-pleated sheets), tertiary and quaternary structure of proteins	
• recognise that enzymes are proteins and describe the characteristics of biological catalysts (enzymes) including that activity depends on the structure and the specificity of the enzyme action	
• recognise that monosaccharides contain either an aldehyde group (aldose) or a ketone group (ketose) and several −OH groups, and have the empirical formula CH_2O	

»	• distinguish between α-glucose and β-glucose, and compare and explain the structural properties of starch (amylose and amylopectin) and cellulose	
	• recognise that triglycerides (lipids) are esters and describe the difference in structure between saturated and unsaturated fatty acids	
	• describe, using equations, the base hydrolysis (saponification) of fats (triglycerides) to produce glycerol and its long chain fatty acid salt (soap), and explain how their cleaning action and solubility in hard water is related to their chemical structure	
	• explain how the properties of polymers depends on their structural features including: the degree of branching in polyethene (LDPE and HDPE), the position of the methyl group in polypropene (syntactic, isotactic and atactic) and polytetrafluorethene.	

Analytical techniques

• explain how proteins can be analysed by chromatography and electrophoresis	
• select and use data from analytical techniques, including mass spectrometry, X-ray crystallography and infrared spectroscopy, to determine the structure of organic molecules	
• analyse data from spectra, including mass spectrometry and infrared spectroscopy, to communicate conceptual understanding, solve problems and make predictions.	

Exam practice

Topic 1: Properties and structure of organic materials

Multiple-choice questions

Solutions start on page 178.

Question 1

Consider the compound shown.

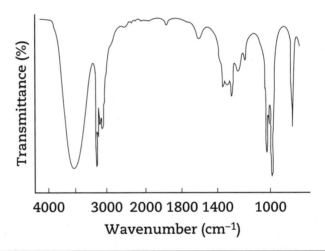

This is an example of

A an amide.

C a carboxylic acid.

B an amine.

D an ester.

Question 2

Analyse the infrared spectrum below to determine the functional groups present and the type of molecule.

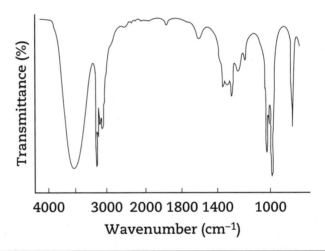

	Functional group	Class of compound
A	C–H stretch and C–H bend	Alkane
B	C–H and C=O	Ester
C	O–H and C–O	Alcohol
D	O–H and C=O	Carboxylic acid

Question 3

Deduce which of the following is an isomer of the compound shown below.

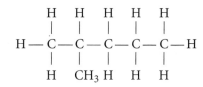

A Heptane

B 2,2-dimethylbutane

C 2,2-dimethylpentane

D 2,2,3-trimethylbutane

Question 4

Identify the option below that gives the correct order of decreasing boiling point.

A 1-Butanol, propanoic acid, butanal, methylbutane

B Butanal, propanoic acid, 1-butanol, methylbutane

C Methylbutane, butanal, 1-butanol, propanoic acid

D Propanoic acid, 1-butanol, butanal, methylbutane

Question 5

Name the product from the reaction between iodine (I_2) and 2-butene.

A 1,1-diiodobutane

B 1,2-diiodobutane

C 2,2-diiodobutane

D 2,3-diiodobutane

Question 6

Isotactic is the term given to a polymer chain that has

A no side groups attached to it.

B side groups attached above and below the chain at regular intervals.

C side groups attached on one side of the chain only.

D side groups randomly attached above and below the chain.

Question 7

When solution X is added to potassium permanganate solution, the colour of the mixture slowly changes from dark purple to colourless. When sodium carbonate solution is added to the mixture, nothing happens.

Deduce which one of the following substances must be solution X.

A 2-methyl-1-pentanol

B 2-methyl-2-pentanol

C 3-methyl-2-pentanol

D 2-methyl-3-pentanone

Question 8

Apply IUPAC rules to name the molecule below.

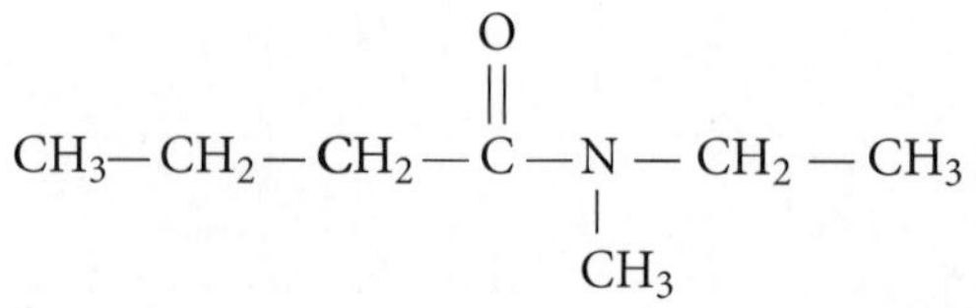

$$CH_3-CH_2-CH_2-\overset{\displaystyle O}{\overset{\displaystyle \|}{C}}-\underset{\displaystyle CH_3}{\overset{\displaystyle |}{N}}-CH_2-CH_3$$

A 3-methylhexylamine

B N-ethyl-N-methylpropanamide

C N-ethyl-N-methylbutanamide

D N-methyl-N-ethylbutanamide

Question 9

When separating a mixture of the amino acids asparagine (Asn) and lysine (Lys) using ion-exchange chromatography, a buffer of pH 7.8 and a cation exchange resin was used. Which one of the following is correct?

	Collected first	Charge	Attraction to stationary phase
A	Asn	+	Attracted
B	Asn	−	Repelled
C	Lys	−	Attracted
D	Lys	+	Repelled

Question 10

Which one of the following is the formula for the nitrile functional group?

A R-CN

B R-CON-R

C R-COO-R

D R-NH$_2$

Question 11

Compound X is known to be an alkane. It is analysed using mass spectrometry and the spectrum below is obtained.

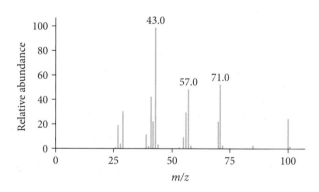

By analysing the spectrum, identify which peak is the base peak.

A 43.0

B 57.0

C 71.0

D 100

Question 12

The sequence of amino acids in a chain is an example of what structure of proteins?

A Primary

B Secondary

C Tertiary

D Quaternary

Question 13

Which of the following substances undergoes a condensation reaction with an acid to form an ester?

A Alcohol

B Alkene

C Amine

D Haloalkane

Question 14

Apply IUPAC rules to name the molecule.

$$CH_3-CH_2-CH_2 \diagdown \qquad \diagup CH_3$$
$$C = C$$
$$CH_3-CH_2 \diagup \qquad \diagdown CH_2-CH_2-CH_2-CH_3$$

A *Cis*-4-methyl-3-propyl-3-octene

B *Cis*-4-ethyl-5-methyl-4-nonene

C *Cis*-5-ethyl-4-methyl-5-nonene

D *Trans*-4-ethyl-5-methyl-4-nonene

Short response questions

Solutions start on page 179.

Question 15 (10 marks)

Compound X contains only carbon and hydrogen. When mixed with chlorine gas in the presence of UV light, compound Q and hydrochloric acid are formed. If some of compound Q is mixed with a sodium hydroxide solution, compound T is produced. Compound T decolourises acidified potassium permanganate solution to produce compound L. When compound L is isolated and mixed with sodium carbonate solution, bubbles of colourless gas are produced.

When a sample of compound Q is mixed with ammonia solution, compound N is produced. If compound N is heated with some of compound L in the presence of concentrated sulfuric acid, compound G is formed. Mass spectral analysis of compound G shows a molecular ion peak at $m/z = 143$.

From this information, draw and apply IUPAC rules to name compounds X, Q, T, L and G, giving your reasons in each case.

Question 16 (7 marks)

A chemist decided to prepare the lipid shown below.

$$
\begin{array}{c}
\text{H} \quad\quad\quad\quad \text{O} \\
\text{H}-\overset{\displaystyle H}{\underset{\displaystyle |}{C}}-O-\overset{\displaystyle \|}{C}-(CH_2)_{16}CH_3 \\
\end{array}
$$

H — C — O — C — (CH$_2$)$_{16}$CH$_3$

H — C — O — C — (CH$_2$)$_{18}$CH$_3$

H — C — O — C — (CH$_2$)$_{16}$CH$_3$

H

a   Draw the structures of the substances required to make this lipid and identify
   the type of reaction that occurs.   4 marks

b   Draw the structures to show how the lipid above could be used to make soap.   3 marks

Question 17 (9 marks)

The structures of three substances belonging to a homologous series are shown below.

A

$CH_3-CH_2-CH_2-\overset{\displaystyle O}{\overset{\displaystyle \|}{C}}-\underset{\displaystyle \underset{\displaystyle CH_3}{|}}{N}-CH_2-CH_3$

B

$CH_3-\overset{\displaystyle O}{\overset{\displaystyle \|}{C}}-\underset{\displaystyle \underset{\displaystyle H}{|}}{N}-CH_2-CH_3$

C

$CH_3-\overset{\displaystyle O}{\overset{\displaystyle \|}{C}}-\underset{\displaystyle \underset{\displaystyle H}{|}}{N}-H$

a   Use IUPAC rules to name each substance.   3 marks

b   For each substance, identify whether it is a primary, secondary or tertiary amide
   and deduce the structure of each R group.   6 marks

Question 18 (4 marks)

The table below gives information about solubility in water and boiling point for some substances.

Substance	Melting point (°C)	Solubility in water (g L^{-1})
1-Butanol	117	73
2-Butanol	98	290
Butanal	74.8	76
Butanone	79	275

a   Explain the differences in melting point and solubility in water between
   1-butanol and 2-butanol.   2 marks

b   Explain the differences in melting point and solubility in water between
   butanal and butanone.   2 marks

Question 19 (6 marks)

Monosaccharides tend to exist in their ring forms, especially in aqueous solution. However, they can be obtained in their open chain form.

Below are the open chain forms of six monosaccharides.

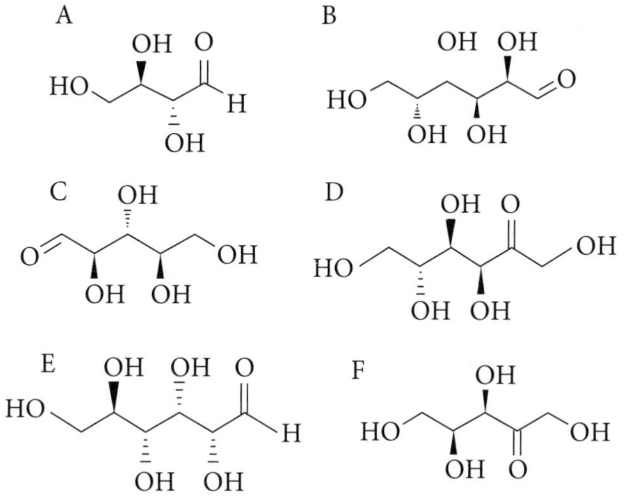

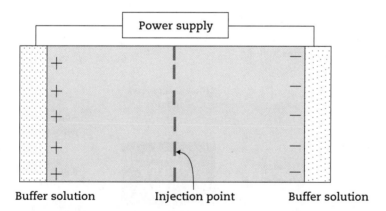

These monosaccharides can be differentiated into *two* distinct types based on their structure. Suggest what this distinction could be and identify each monosaccharide as being one or the other.

Question 20 (6 marks)

Amino acids can be separated and characterised using electrophoresis. A mixture of threonine and aspartic acid is added to a buffer solution of pH 4. This mixture is injected into one of the wells as shown in the diagram.

a   Explain how the mixture of amino acids is separated using this method.   3 marks

b   On the diagram, indicate where you think the bands for threonine and aspartic acid will settle.

Explain your reasoning in each case.   3 marks

Chapter 5
Topic 2: Chemical synthesis and design

Topic summary

This topic explores the business of chemistry, where a minute change in a chemical process can cost millions of dollars. In the real world, away from the laboratory and theoretical chemistry, the business of chemistry is constantly changing and evolving. However, these changes are informed by the progress made through research and development in laboratories. Industrial processes involving an understanding of chemical equilibrium are investigated. Specific reactions designed to address environmental concerns are studied; these include enzymes to produce biodiesel and fuel cells that use hydrogen. The environmental impact of our irresponsible reliance on synthetic processes is addressed, along with attempts to resolve these issues in economically viable ways. Finally, the production of synthetic polymers is compared with natural polymers. The subject of molecular manufacturing is studied, with its focus on the design and construction of specific substances at the molecular level and the benefits that this can bring.

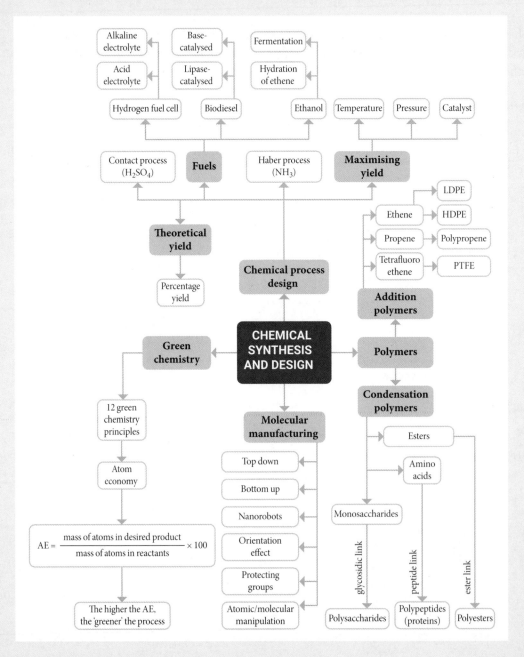

5.1 Chemical synthesis

There are often many ways to use **chemical synthesis** to create a product with specific properties. Once the desired product is identified, reactants and a **reaction pathway** are developed that consider economic viability, safety and environmental factors such as wastage.

5.1.1 Theoretical yield

To be economically viable, a process must produce the product as quickly as possible with the minimum cost. In other words, the maximum amount of product must be produced as quickly as possible.

Before factors that will maximise the yield of a reaction can be considered, it is important to understand the concept of **theoretical yield**.

Calculating theoretical yield

This is the amount of product that can theoretically be produced from reactants.

Worked example

Question: The ester pentyl ethanoate, an ester with a banana smell, can be produced by heating a mixture of 1-pentanol and ethanoic acid together with a small amount of concentrated sulfuric acid according to the equation:

$$CH_3CH_2CH_2CH_2CH_2OH + CH_3COOH \rightarrow CH_3COOCH_2CH_2CH_2CH_2CH_3$$
$$\text{1-pentanol} \qquad \text{ethanoic acid} \qquad \text{pentyl ethanoate}$$

If 3.98 g of 1-pentanol is mixed with 2.46 g of ethanoic acid, what mass of the ester will be produced?

Step 1

Calculate the number of moles of both reactants.

$$CH_3CH_2CH_2CH_2CH_2OH + CH_3COOH \rightarrow CH_3COOCH_2CH_2CH_2CH_2CH_3$$

$$\frac{3.98}{88.15} = 0.0451 \text{ mol 1-pentanol} \qquad \frac{2.46}{60.05} = 0.0410 \text{ mol ethanoic acid}$$

Step 2

Determine the limiting reagent.

The **limiting reagent** in a reaction is the reactant that is used up completely. The reactant left over is the **excess reagent**.

As this is a 1:1 ratio, the ethanoic acid is the limiting reactant.

Step 3

Use the limiting reagent to determine the theoretical mass of the product.

The limiting reagent and the product are in a 1:1 mole ratio.

Therefore, number of moles of product = number of moles limiting reagent.

Mass of product = 0.0410 mol pentyl ethanoate $\times$ 130.19 g mol^{-1}

= 5.34 g pentyl ethanoate theoretically produced.

Calculating percentage yield

These calculations are carried out when a mass of product is given and compared with the theoretical yield.

Worked example

Question: A 1.94 g sample of ethene reacts with an excess of chlorine to produce 5.78 g of 1,2-dichloroethane according to the equation:

$$C_2H_4(g) + Cl_2(g) \rightarrow C_2H_4Cl_2(l)$$

Calculate the percentage yield for this reaction.

Step 1

Calculate the amount of product that can be theoretically produced from the given mass of reactants.

$$C_2H_4(g) + Cl_2(g) \rightarrow C_2H_4Cl_2(l)$$

$$\frac{1.94}{28.05} = 0.0692 \text{ mol} \rightarrow 0.0692 \text{ mol}$$

Mass of product = **0.0692 × 98.96 = 6.85 g**

Step 2

Calculate the percentage yield using the theoretical mass calculated in step 1 and the actual mass given in the question.

> **Hint**
>
> Use the formula for percentage yield given on page 1 of the *Chemistry formula and data booklet*.
>
> $$= \frac{\text{Experimental yield (or actual yield)}}{\text{Theoretical yield}} \times \frac{100}{1}$$
>
> $$= \frac{5.78}{6.85} \times 100 = 84.4\%$$

5.1.2 Designing optimal processes

The Haber process

The **Haber process** is the main method for the industrial production of ammonia, NH_3. Ammonia is an extremely important compound necessary for the production of fertilisers, dyes, explosives, and a number of industrial and household cleaning products.

The Haber process was designed to maximise the amount of ammonia produced whilst minimising the production of environmental pollutants, and is represented by the following equation:

$$N_2(g) + 3H_2(g) \rightleftharpoons 2NH_3(g) \quad \Delta H = -92.4 \text{ kJ mol}^{-1}$$

The economically important issues here are amount produced (yield) and speed of production (rate). These requirements determine the optimum operating conditions of temperature and pressure.

Maximising yield

When choosing the optimum pressures and temperatures, it is important to remember Le Chatelier's principle.

> **Hint**
>
> **Key concept**
>
> Le Chatelier's principle:
>
> If a system at equilibrium is subject to a change in conditions, then the system will behave in such a way as to partially counteract the change.

Pressure

Inspection of the equation shows that on the left-hand side of the equation there is a total of **four** molecules of gas (1 mol N_2 + 3 mol H_2). On the right-hand side of the equation there is only **two** molecules of gas.

Therefore:

$$N_2(g) + 3H_2(g) \rightleftharpoons 2NH_3(g)$$
$$\text{high pressure} \qquad \text{low pressure}$$

Given that the amount of NH_3 needs to be maximised, the system needs to shift as far as possible to the *right*.

If the pressure on the system is *increased*, the system will shift to the side that will remove the pressure, the low pressure side. Therefore, one way to increase yield in this instance would be to increase the pressure in the reaction vessel, causing the reaction to shift to produce more NH_3.

Temperature

Inspection of the thermochemical equation:

$$N_2(g) + 3H_2(g) \rightleftharpoons 2NH_3(g) \ \Delta H = -92.4 \text{ kJ mol}^{-1}$$

The forward reaction has a negative ΔH value; it is an **exothermic reaction**, it produces heat. Therefore, the reverse reaction is an **endothermic reaction**, it absorbs heat.

According to Le Chatelier's principle, in order to shift the system to the right (keeping in mind the goal is to produce the most NH_3), heat needs to be removed. This will force the system to shift right in order to produce more heat and replace the heat that was taken away.

Therefore, the temperature of the system needs to be *decreased*.

BUT

At low temperatures, the yield may be high, but the reaction will be very slow. A compromise needs to be made between yield and rate.

Generally, the operating conditions for the Haber process are:

- Pressure = 200 atmospheres (atm)
- Temperature = 400–450°C
- Catalyst = iron oxide.

The contact process

The contact process is the method by which sulfuric acid is produced. Sulfuric acid is one of the most important chemicals produced in the world today. It is used in the production of metals, dyes, paints, fertilisers, detergents and fibres.

Sulfuric acid is produced in three main stages.

Stage 1

Sulfur is burnt in air to produce sulfur dioxide.

$$S(s) + O_2(g) \rightarrow SO_2(g)$$

Stage 2

The $SO_2(g)$ produced in stage 1 is reacted with $O_2(g)$ in the presence of vanadium pentoxide (V_2O_5) catalyst to produce sulfur trioxide, $SO_3(g)$.

$$2SO_2(g) + O_2(g) \rightleftharpoons 2SO_3(g)$$

Stage 3

The $SO_3(g)$ produced in stage 2 is added to concentrated sulfuric acid.

$$SO_3(g) + H_2SO_4(l) \rightarrow H_2S_2O_7(l)$$

$H_2S_2O_7$ is known as oleum, or fuming sulfuric acid.

This is then safely reacted with water to give concentrated sulfuric acid.

$$H_2S_2O_7(l) + H_2O(l) \rightarrow 2H_2SO_4(l)$$

Maximising yield

The most important stage in the contact process is stage 2. The thermochemical equation for this reaction is:

$$2SO_2(g) + O_2(g) \rightleftharpoons 2SO_3(g) \quad \Delta H = -196\,kJ\,mol^{-1}$$

Pressure

High pressure is used because this forces the reaction to shift to the side with fewer gas molecules.

However, the pressure is only a few atmospheres because of the unacceptably high economic cost of operating the process at very high pressures.

Temperature

The thermochemical equation shows that the combustion of SO_2 is an exothermic reaction. Therefore, a low temperature would favour the products but, as mentioned previously with the Haber process, a compromise between rate and yield must be made.

Generally, the operating conditions for the contact process are:

- Pressure = About 2 atmospheres (atm)
- Temperature = 400–450°C
- Catalyst = V_2O_5

5.1.3 Synthesis of fuels

The world's reliance on fossil fuels as the primary source of energy is diminishing reserves and contributing to climate change. Therefore, in recent years, a great deal of effort has gone into developing alternative sources of fuel. Two of the most promising are the **biofuels** ethanol and hydrogen.

Synthesis of ethanol

The two most important methods for creating ethanol are fermentation and hydration of ethene.

Fermentation

A suitable grain or fruit is mashed up with water and mixed with a yeast strain. The glucose produced is then converted to ethanol:

$$C_6H_{12}O_6(aq) \rightarrow 2C_2H_5OH(aq) + 2CO_2(g)$$

Hydration of ethene

This is the most common method of producing ethanol, due to the low cost of ethene produced from the **cracking** of crude oil. However, ethanol produced this way isn't really a *bio*fuel because it does not come from a crop. It comes from a fossil fuel – crude oil!

$$CH_2{=}CH_2(g) + H_2O(g) \rightarrow CH_3CH_2OH(l)$$

The conditions for this process are about 60–70 atm and 300°C with a phosphoric acid catalyst.

Synthesis of biodiesel

Biodiesel is a mixture of alkyl esters of fatty acids and can be produced from any vegetable oil or animal fat. It is produced by two main processes.

Base-catalysed method

Diesel from crude oil is fuel with a hydrocarbon chain of 8–21 carbon atoms. Biodiesel is similar, but it uses **triglycerides**, an ester that comes from animal or vegetable fats. This process is called **transesterification**, it combines the triglyceride and a cheap alcohol, such as methanol, and produces a new ester (biodiesel) and a new alcohol.

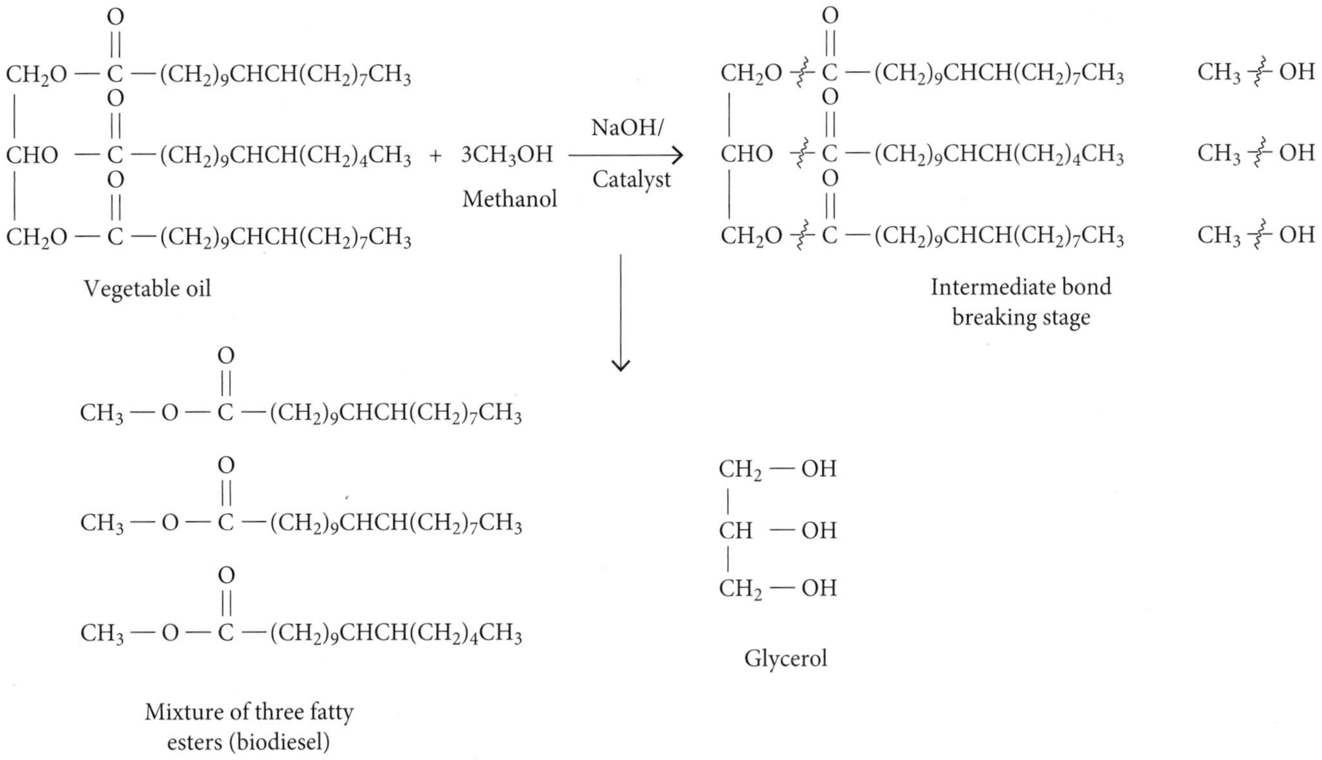

$$CH_2O - \overset{\overset{\displaystyle O}{||}}{C} - (CH_2)_9CHCH(CH_2)_7CH_3$$
$$CHO - \overset{\overset{\displaystyle O}{||}}{C} - (CH_2)_9CHCH(CH_2)_4CH_3 \;+\; 3CH_3OH \;\xrightarrow[\text{Catalyst}]{\text{NaOH/}}\;$$
$$CH_2O - \overset{\overset{\displaystyle O}{||}}{C} - (CH_2)_9CHCH(CH_2)_7CH_3$$

Vegetable oil

Methanol

$$CH_2O \nmid \overset{\overset{\displaystyle O}{||}}{C} - (CH_2)_9CHCH(CH_2)_7CH_3 \qquad CH_3 \nmid OH$$
$$CHO \nmid \overset{\overset{\displaystyle O}{||}}{C} - (CH_2)_9CHCH(CH_2)_4CH_3 \qquad CH_3 \nmid OH$$
$$CH_2O \nmid \overset{\overset{\displaystyle O}{||}}{C} - (CH_2)_9CHCH(CH_2)_7CH_3 \qquad CH_3 \nmid OH$$

Intermediate bond
breaking stage

$$CH_3 - O - \overset{\overset{\displaystyle O}{||}}{C} - (CH_2)_9CHCH(CH_2)_7CH_3$$
$$CH_3 - O - \overset{\overset{\displaystyle O}{||}}{C} - (CH_2)_9CHCH(CH_2)_7CH_3$$
$$CH_3 - O - \overset{\overset{\displaystyle O}{||}}{C} - (CH_2)_9CHCH(CH_2)_4CH_3$$

Mixture of three fatty
esters (biodiesel)

$$CH_2 - OH$$
$$CH \; - OH$$
$$CH_2 - OH$$

Glycerol

FIGURE 5.1 The process of transesterification

Biodiesel synthesis breaks the bonds between the glycerol and fatty acids and adds a CH_3 to the end of the fatty acid.

To maximise the yield of biodiesel, a base catalyst is used, usually sodium or potassium hydroxide. The issue here is that soap can be produced, which affects the yield. A way of reducing this risk is to use lipase as a catalyst.

Hydrogen in fuel cells

The alkaline fuel cell was discussed at length in Chapter 2 but it is worth remembering that instead of converting the chemical potential energy in the bonds in hydrogen and oxygen into heat, a fuel cell converts the chemical energy directly into electrical energy.

Generally, the operation of all hydrogen fuel cells is very similar but different fuel cells can differ in the type of electrolyte used. The two most used types of electrolyte are acid, such as phosphoric acid, and alkaline, such as potassium hydroxide.

Acid electrolyte

The reactions at the electrodes are:

Anode (negative electrode): $2H_2(g) \rightarrow 4H^+(aq) + 4e^-$ $E° = +0.00$ V
Cathode (positive electrode): $O_2(g) + 4H^+(aq) + 4e^- \rightarrow 2H_2O(l)$ $E° = +1.23$ V

Alkaline electrolyte

The reactions at the electrodes are:

Anode: $O_2(g) + 2H_2O(l) + 4e^- \rightarrow 4OH^-(aq)$ $E° = +0.40$ V
Cathode: $2H_2(g) + 4OH^-(aq) \rightarrow 4H_2O(l) + 4e^-$ $E° = -0.83$ V
For a detailed description of the operation of an alkaline fuel cell, see Chapter 2.

5.2   Green chemistry

5.2.1 The principles of green chemistry

In the 1990s, the 12 principles of green chemistry were developed. These principles can be summarised as:

1 Prevent waste
2 Maximise atom economy
3 Design less hazardous chemical syntheses
4 Design safer chemicals and products
5 Use safer solvents and reaction conditions
6 Increase energy efficiency
7 Use renewable reactants
8 Avoid using chemical derivatives
9 Use catalysts
10 Design chemicals that will degrade
11 Prevent pollution
12 Minimise the potentials of chemical accidents

5.2.2 Atom economy

Atom economy relates to the efficiency of a chemical process at converting the mass of reactants into desired products and is calculated using the equation:

$$\text{Atom economy} = \frac{\text{Mass of atoms in desired product}}{\text{Mass of atoms in the reactants}} \times 100$$

The greater the atom economy, the less wastage of atoms and the greener the chemistry.

Worked example

Question: Iron is produced in an industrial process by the reduction of iron(III) oxide by carbon monoxide according to the equation:

$$Fe_2O_3(s) + 3CO(g) \rightarrow 2Fe(l) + 3CO_2(g)$$

Calculate the atom economy for this process.

Step 1

Calculate the total molar mass of the reactants.

$$M = (Fe_2O_3) + (3 \times CO) = (159.70) + (3 \times 28.01) = \textbf{243.73 g mol}^{-1}$$

Step 2

Calculate the total mass of the *desired* product.

$$M = (2 \times Fe) = (2 \times 55.85) = \textbf{111.7 g mol}^{-1}$$

Step 3

Calculate the atom economy.

$$\text{Atom economy} = \frac{111.7}{243.73} \times 100 = 45.83\%$$

5.3 Macromolecules: polymers, proteins and carbohydrates

5.3.1 Describing the production of addition polymers

Addition polymers form by adding molecules together without the loss of any atoms. They are generally formed from ethene or substituted alkenes. The double bond is ideal for this because one of the bonds in the double bond can break to form a radical at both ends of the monomer molecule. The radical can then form bonds with other monomer molecules.

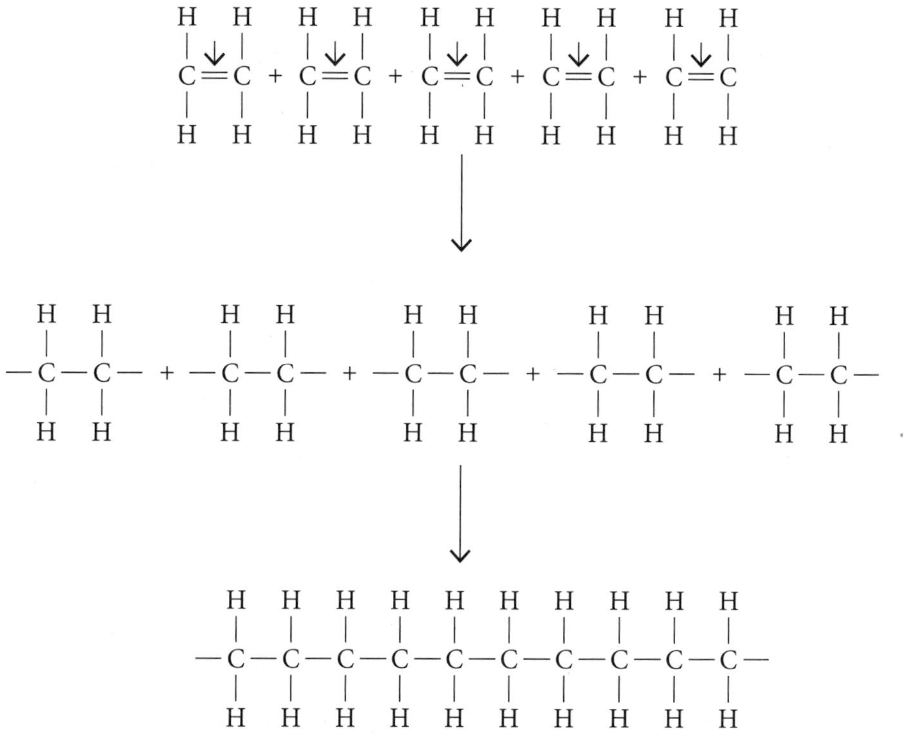

FIGURE 5.2 The production of addition polymers

There are two main types of polyethene.

LDPE (Low density):

$$CH_2 = CH_2 \xrightarrow[\substack{\text{High pressure (1000–3000 atm)} \\ \text{Initiator}}]{\text{High temp (300°C)}} - (CH_2 - CH_2)_n - (BRANCHED)$$

HDPE (High density):

$$CH_2 = CH_2 \xrightarrow[\substack{\text{Low pressure (10–50 atm)} \\ \text{Ziegler-Natta catalyst}}]{\text{Low temp (60°C)}} - (CH_2 - CH_2)_n - (UNBRANCHED)$$

Other important addition polymers include polypropene (PP plastic) and polytetrafluoroethene (PTFE, known as Teflon™) and have been discussed in detail in Chapter 3.

5.3.2 Describing the production of condensation polymers

Condensation polymers form when two monomer molecules combine to eliminate a small molecule, usually water – hence the name condensation polymers.

There are two types of condensation polymer:

Synthetic condensation polymers – polyesters, polyamides

Natural condensation polymers – polysaccharides (cellulose and starch) and proteins

Polyesters

Polyesters can be produced from:

1   A single monomer such as 2-hydroxypropanoic acid (lactic acid). This can polymerise with other molecules via an esterification reaction with the carboxyl group at one end of the molecule and the hydroxyl group of an adjacent monomer.

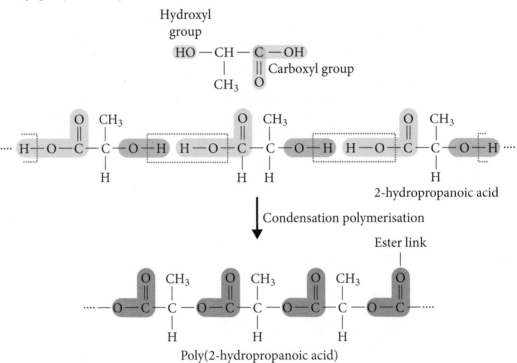

FIGURE 5.3 The formation of a polyester from a single monomer

2   Two different monomers, a diol and a dicarboxylic acid:

HO—CH$_2$—CH$_2$—OH

1, 2-ethanediol
(ethylene glycol)

Terephthalic acid

Polyethylene terephthalate (PET)

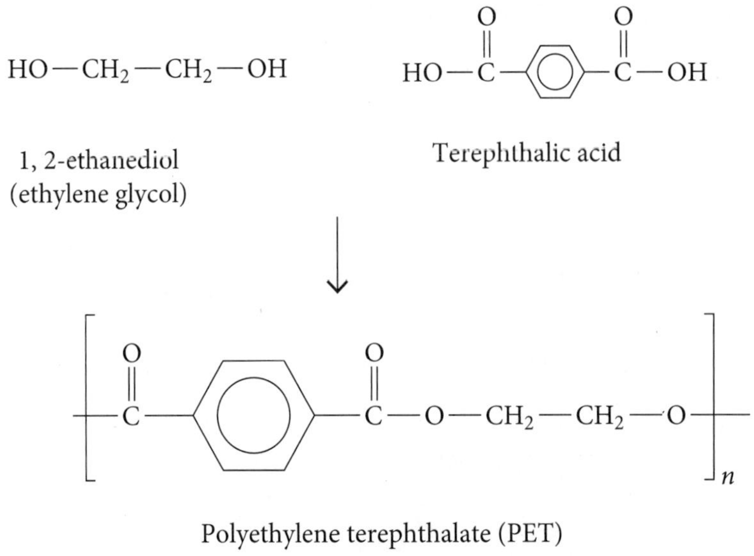

FIGURE 5.4 Two different monomers, a diol and a dicarboxylic acid

Polypeptides (proteins)

Polypeptides form when amino acid monomers join.

The simplest amino acid is glycine. Like all amino acids it contains a carboxyl group and an amino group:

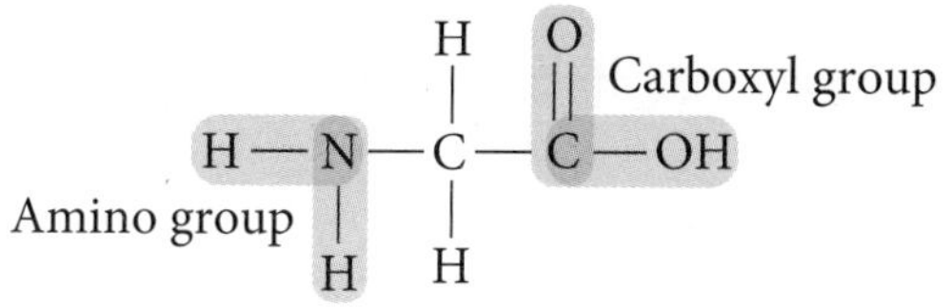

FIGURE 5.5 The formation of glycine

When amino acids join, they eliminate a water molecule in a condensation reaction:

Peptide link

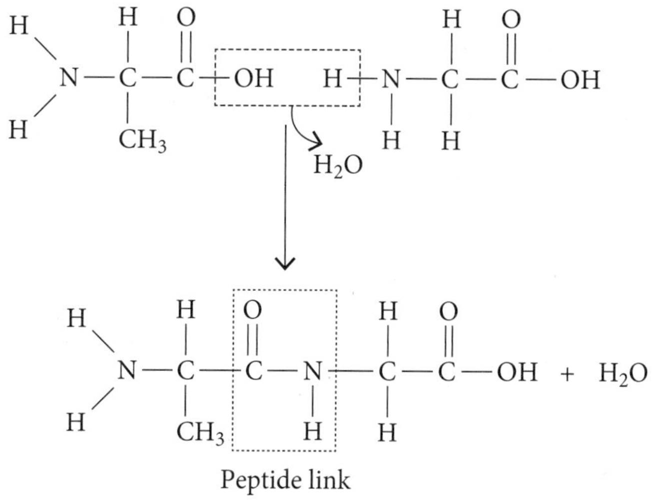

FIGURE 5.6 The dipeptide alanylglycine

This dipeptide, alanylglycine, has the carboxyl and hydroxyl functional groups at either end of the molecule, which can continue to build up the polypeptide chain.

Polysaccharides

Polysaccharides form when monosaccharide monomers join in a condensation reaction.
The link that joins two monosaccharides together is called a **glycosidic link (or bond)**.

5.3.3 Advantages and disadvantages of polymer use

It is important to consider the structural features of polymers when selecting one for a particular purpose.

Crystallinity

Long polymer chains can move close together, resulting in stronger intermolecular forces that lead to regions of greater rigidity and higher softening and melting points. These regions are called **crystalline** regions. They also have greater density, opaque appearance and are impermeable to air, water and other chemicals. Regions where the polymer chains are disordered and tangled are called **amorphous** regions. These regions are less dense, have lower softening and melting points, are transparent and are permeable to air, water and other chemicals. The crystallinity of a polymer can be controlled according to the specifications required.

Branching

Chains with little or no branching are denser, less transparent and less flexible than those with more branching.

Chain length

The longer the chain length and the smaller the differences in chain length of the polymer chains, the higher the melting point and the harder the polymer.

Side groups

Large side groups reduce the flexibility of a polymer.

Cross-linking

The hardness and rigidity of a polymer can be increased by adding cross-linking agents into the manufacturing process.

For example, natural or synthetic rubber is soft and pliable. If the rubber is heated with sulfur, S–S bonds form between the polymer chains, making the rubber much harder and less flexible, suitable for use in tyres. This process is called vulcanisation.

Stability and biodegradability

The strong covalent –C–C– bonds are quite stable and do not degrade easily. Most synthetic polymers mostly containing these bonds are *not* biodegradable.

The biodegradability of synthetic polymers can be increased by copolymerising them with segments of natural polymers such as starch molecules.

Disadvantages of polymer use

The two main disadvantages with polymer use are environmental concerns around their manufacture and disposal.

Manufacturing plastics

In recent decades, society has become increasingly dependent on plastics. The supply of natural resources such as crude oil, coal and natural gas is finite. There is a great deal of research being carried out to find alternative sources of the raw materials required for plastics production.

Disposal of plastics

Given the huge amount and variety of plastics in use, a great effort must be made to recycle them. The rewards are substantial. Less energy is required to recycle plastics than it is to make them from scratch and less raw material (from an already dwindling supply) is needed.

Evidence is coming to light that when plastics are in the environment, they begin to fragment due to the action of sunlight, chemicals and even microorganisms. It is unclear how these polymer fragments will affect the environment and ultimately, us.

5.3.4 Condensation reactions of amino acids

Forming the peptide bond

As mentioned previously, all amino acids contain an amino group and a carboxyl group, both of which form links (or bonds) with other amino acids. These bonds are called peptide bonds.

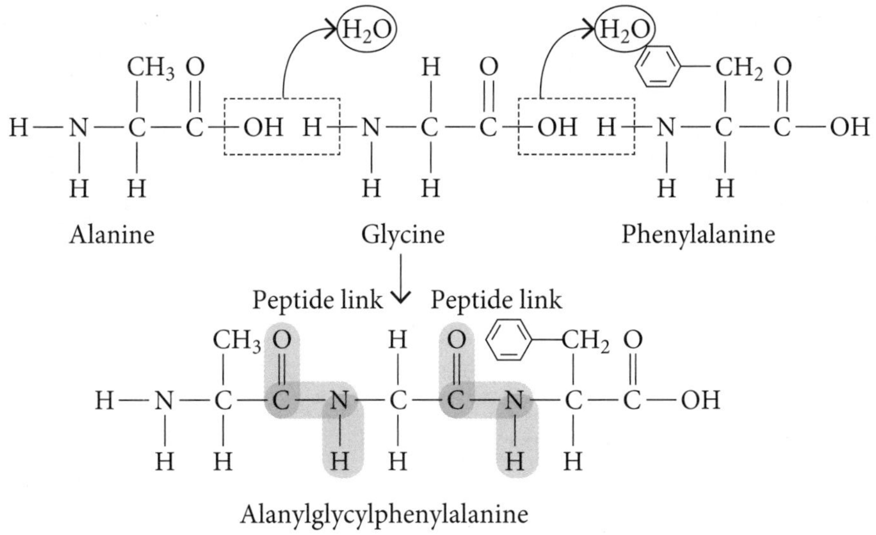

FIGURE 5.7 Formation of peptide bonds

5.3.5 Condensation reactions of monosaccharides

Disaccharides

When two monosaccharides join, they form a **disaccharide**. This is a condensation reaction. Important disaccharides include sucrose, maltose and lactose.

Lactose is formed from glucose and galactose.

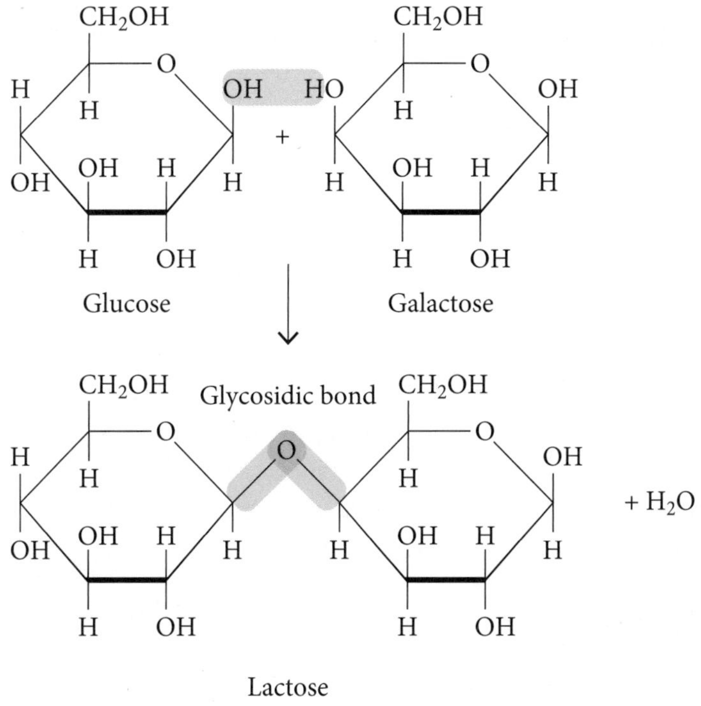

FIGURE 5.8 The formation of lactose

Disaccharides are large molecules, too large to pass through cell membranes. They are broken down to their constituent monomers in the digestive system. The reaction to produce the disaccharide was a condensation reaction. The reaction to break the disaccharide is a hydrolysis reaction.

Polysaccharides

Polysaccharides are all polymers of glucose with different glycosidic links, amounts of branching and degrees of polymerisation.

The starch polymer consists of two parts. The smaller part (about 20 per cent), amylose, is a straight-chain polymer that can form spirals due to intrachain hydrogen bonding. Amylopectin, which forms the major part (80 per cent), is a highly branched chain. These chains occur when one glucose molecule condenses with three other glucose molecules instead of the normal two. This part of the molecule cannot form spirals.

Cellulose has unbranched molecules that preferentially form hydrogen bonds with each other to create a strong, cross-linked structure. This cross-linking gives cellulose its strength and rigidity. Unlike starch, cellulose is insoluble in water. Cellulose is a highly crystalline polysaccharide, which accounts for many of its properties.

5.4 Molecular manufacturing

Molecular manufacturing is the production of molecules, crystals and other structures that have been designed for a specific purpose and to have a specific shape or composition.

Molecular manufacturing can be done in three ways, as outlined below.

5.4.1 The orientation effect (mechanosynthesis)

This involves positioning molecules so that the desired functional groups line up. This ensures that reaction processes are much more efficient and produce less waste and fewer undesired products. This occurs naturally in protein synthesis and DNA replication. The difficulty is, for it to work effectively, every atom in the vicinity of the reaction taking place is affected. This interferes with the reaction taking place, making the process less efficient.

5.4.2 Manipulation of structures at the atomic or molecular level

Structures can be manipulated at the atomic or molecular level to control the properties of the final substance. Bonds between atoms, ions or molecules can be altered or specific atoms can be added or substituted.

Use of protecting groups

Protecting groups can be added to a substance to inhibit certain reactive sites from reacting, thereby increasing the specificity of the substance.

Top-down and bottom-up approaches

A top-down approach to molecular manufacturing involves starting with a larger substance and removing parts of it until a smaller, more useful substance is left.

A bottom-up approach involves adding atoms and molecules together to make a larger, more complex molecule.

5.4.3 Protein synthesis

The medical and pharmaceutical industries are working on techniques for artificial protein synthesis, which have enormous potential for high specificity drugs and artificial tissue.

Joining amino acids

As discussed previously, amino acids join in a condensation reaction. However, specific proteins require a very precise sequence of amino acids, which does not come about by chance (and proteins are massive, whereas peptides we make in the lab are a fraction of that size).

For a protein to form, the two functional groups of the constituent amino acids must line up perfectly.

The blueprint for every protein in a living organism is found in the DNA in its cells. This blueprint is copied and moved out of the nucleus of the cell by mRNA and tRNA. This is facilitated by a ribosome, which is a specialised organelle. The job of the ribosome is to ensure that each amino acid is in the correct place, so that its functional groups are aligned. This enables amino acids to join and create a polypeptide chain.

This is the natural way to synthesise proteins. Aspects of this process have been copied in order to carry out artificial protein synthesis. One method uses a resin onto which amino acids are attached so that they are in a particular orientation. More amino acids are attached until a protein is formed, after which the resin is removed. In this case, the resin acts like a ribosome. Another method uses a molecular machine very similar to a ribosome in its structure and function.

5.4.4 Carbon nanotubes

Carbon can exist in many forms, diamond and graphite being the most familiar. However, it is possible to produce sheets of covalently bonded carbon atoms, just one atom thick. These sheets can be rolled into **nanotubes**, which then can be used in applications, including nanocircuitry, as molecular wires and very strong nanocables.

5.4.5 Nanorobots

Nanorobots are artificially created proteins designed to perform specific mechanical tasks. A molecular robot can be used to recognise a particular shape or structure, such as diseased cells or viruses in the body.

5.4.6 Chemical sensors

A **biosensor** is a sensor that detects a change in a biological component, such as a particular chemical. A **nanosensor** is a biosensor that has been miniaturised down to the nanoscale. Promising fields of research include using nanosensors to detect propanone (a compound produced by people with diabetes) in the breath. Nanosensors can detect chemicals produced by cancer cells before symptoms present.

Glossary

addition polymers
polymers formed by adding together molecules
without the loss of any atoms

amorphous
without a clearly defined shape or form

atom economy
the efficiency of a chemical reaction. The greatest
efficiency is where the greatest percentage of atoms in
the reactant end up in the product

biofuel
fuels produced from plants, algae or animal waste

biosensor
a structure that can detect a change in a biological
component

chemical synthesis
to create a product with specific properties artificially
rather than find it occurring naturally

condensation polymer
when two monomer molecules join to eliminate a
small molecule (often water)

cracking
the process of heating larger organic molecules and
converting them into smaller, more useful molecules

crystalline
having a regular, ordered arrangement of atoms

disaccharide
a sugar made from two monosaccharide units

endothermic reaction
a reaction in which the products have more enthalpy
than the reactants, meaning that energy will be
absorbed from the surroundings

excess reagent
the reactant that remains after a reaction is complete

exothermic reaction
a reaction in which the reactants have more enthalpy
than the products and therefore energy will be released
to the surroundings

glycosidic link (or bond)
the link that joins two monosaccharides together

Haber process
a process used in the industrial production of ammonia

A+ DIGITAL FLASHCARDS
Revise this topic's key terms
and concepts by scanning
the QR code or typing the
URL into your browser.

https://get.ga/aplus-
qce-chem-u34

limiting reagent
the reactant that is completely used up during a
reaction

molecular manufacturing
the production of molecules, crystals and other
chemical structures, that have been designed to have
a specific shape or chemical composition

polypeptide
a polymer formed when amino acid monomers join

polysaccharide
a carbohydrate molecule made from many
monosaccharide units

nanorobots
artificially created proteins designed to perform
specific mechanical tasks

nanosensor
a sensor that is nano-sized

nanotubes
a tube constructed on the nanoscale; most commonly
made from carbon

reaction pathway
a series of reactions leading to a specific outcome or
product

theoretical yield
the amount of product that can theoretically be
produced from reactants

transesterification
the process of converting one ester into another, for
example, in the industrial production of biodiesel,
converting a lipid into methyl esters of fatty acids, and
by replacing the glycerol with methanol

triglyceride
an ester formed from glycerol and three fatty acid
molecules

Revision summary

Use the following summary of syllabus dot points and key knowledge within Unit 4 Topic 2 to ensure that you have thoroughly reviewed the content. Provide a brief definition or comment for each item to demonstrate your understanding or code them using the traffic light system – Green (all good); Amber (needs some review); Red (priority area to review).

Chemical synthesis	
• appreciate that chemical synthesis involves the selection of particular reagents to form a product with specific properties	
• understand that reagents and reaction conditions are chosen to optimise the yield and rate for chemical synthesis processes, including the production of ammonia (Haber process), sulfuric acid (contact process) and biodiesel (base-catalysed and lipase-catalysed methods)	
• understand that fuels, including biodiesel, ethanol and hydrogen, can be synthesised from a range of chemical reactions including addition, oxidation and esterification	
• understand that enzymes can be used on an industrial scale for chemical synthesis to achieve an economically viable rate, including fermentation to produce ethanol and lipase-catalysed transesterification to produce biodiesel	
• describe, using equations, the production of ethanol from fermentation and the hydration of ethene	
• describe, using equations, the transesterification of triglycerides to produce biodiesel	>>

>>	• discuss, using diagrams and relevant half-equations, the operation of a hydrogen fuel cell under acidic and alkaline conditions.	
	• calculate the yield of chemical synthesis reactions by comparing stoichiometric quantities with actual quantities and by determining limiting reagents.	
Green chemistry		
	• appreciate that green chemistry principles include the design of chemical synthesis processes that use renewable raw materials, limit the use of potentially harmful solvents and minimise the amount of unwanted products	
	• outline the principles of green chemistry and recognise that the higher the atom economy, the 'greener' the process	
	• calculate atom economy and draw conclusions about the economic and environmental impact of chemical synthesis processes.	
Macromolecules: polymers, proteins and carbohydrates		
	• describe, using equations, how addition polymers can be produced from their monomers including polyethene (LDPE and HDPE), polypropene and polytetrafluorethene	
	• describe, using equations, how condensation polymers, including polypeptides (proteins), polysaccharides (carbohydrates) and polyesters can be produced from their monomers	>>

» • discuss the advantages and disadvantages of polymer use, including strength, density, lack of reactivity, use of natural resources and biodegradability	
• describe the condensation reaction of 2-amino acids to form polypeptides (involving up to three amino acids), and understand that polypeptides (proteins) are formed when amino acid monomers are joined by peptide bonds	
• describe the condensation reaction of monosaccharides to form disaccharides (lactose, maltose and sucrose) and polysaccharides (starch, glycogen and cellulose), and understand that polysaccharides are formed when monosaccharides monomers are joined by glycosidic bonds.	
Molecular manufacturing	
• appreciate that molecular manufacturing processes involve the positioning of molecules to facilitate a specific chemical reaction; such methods have the potential to synthesise specialised products, including proteins, carbon nanotubes, nanorobots and chemical sensors used in medicine.	

Exam practice

Topic 2: Chemical synthesis and design

Multiple-choice questions

Solutions start on page 183.

Question 1

After a chemical reaction has occurred, the reactant that is remaining is called the

A equivalent reagent.

B excess reagent.

C limiting reagent.

D stoichiometric reagent.

Question 2

The compound 1-iodobutane can be prepared by mixing butane with iodine vapour in the presence of UV light, according to the equation:

$$C_4H_{10}(g) + I_2(g) \rightarrow C_4H_9I(l) + HI(g)$$

Calculate the atom economy of the process with iodobutane as the desired product.

A 41%

B 49.73%

C 58.99%

D 99.45%

Question 3

Which two functional groups are required to form a polyester?

A Alcohol and aldehyde

B Alcohol and amine

C Alcohol and carboxylic acid

D Amine and carboxylic acid

Question 4

L and D forms of amino acids exhibit chirality. This means that they

A are mirror images of each other.

B are structural isomers.

C are stereoisomers.

D have different R groups attached to the α-carbon atom.

Question 5

The most common way of producing biodiesel is by

A base-catalysed transesterification of vegetable or animal fat.

B hydration of ethene.

C lipase-catalysed transesterification of vegetable or animal fat.

D reaction of vegetable or animal fat with oxygen.

Question 6

Maximising atom economy is desirable because

A fewer atoms are undesirable products.

B it is quicker.

C more atoms are left over at the end of the process.

D more reactants can be recycled.

Question 7

Identify the type of bond that links monomer molecules in a polysaccharide.

A Amide

B Ester

C Glycosidic

D Peptide

Question 8

In an alkaline hydrogen fuel cell, electricity is generated when

A $O_2(g)$ enters the cell and is oxidised at the anode, thereby producing electrons.

B $H_2(g)$ enters the cell and is oxidised at the cathode, thereby producing electrons.

C $H_2(g)$ enters the cell and is reduced at the cathode, thereby accepting electrons.

D $O_2(g)$ enters the cell and is reduced at the cathode, thereby accepting electrons.

Question 9

Glycogen is an example of a

A dipeptide.

B polyester.

C polypeptide.

D polysaccharide.

Question 10

Sulfuric acid is produced in the

A Bosch process.

B contact process.

C Haber process.

D Solvay process.

Question 11

The orientation effect is also referred to as

A chemical kinetics.

B mechanosynthesis.

C retrosynthesis.

D top-down approach.

Question 12

The industrial production of ammonia is achieved through the Haber process, represented by the thermochemical equation:

$$N_2(g) + 3H_2(g) \rightleftharpoons 2NH_3(g) \; \Delta H = -92.4 \text{ kJ mol}^{-1}$$

Which set of conditions would maximise the production of ammonia?

	Pressure	Temperature
A	high	high
B	high	low
C	low	high
D	low	low

Question 13

The use of catalysts can help make a reaction meet the aims of green chemistry by

A enabling lower temperatures and pressures to be used.

B increasing atom economy.

C minimising the risk of accidents.

D producing fewer hazardous materials.

Question 14

Which one of the following best represents a peptide link?

A

$$-C-N-$$
with O double-bonded to C (above) and H below N

B

$$-C-O-$$
with O double-bonded to C (below)

C

$$-C-O-C-$$

D

$$-C-C\equiv N$$
with H above and H below the first C

Short response questions

Solutions start on page 184.

Question 15 (6 marks)

a Draw the dipeptide formed when alanine (Ala) and valine (Val) are joined in a condensation reaction.   3 marks

b On your diagram, identify the:

　　i type of bond between the original monomers.

　　ii position and name of the functional groups present.

　　iii name of the dipeptide.   3 marks

Question 16 (6 marks)

Ethanol can be produced in industrial quantities by two main methods.

Fermentation:

$$C_6H_{12}O_6(aq) \rightarrow 2C_2H_5OH(aq) + 2CO_2(g)$$

Hydration of ethene:

$$C_2H_4(g) + H_2O(aq) \rightarrow C_2H_5OH(l)$$

a   Calculate the atom economy for each process.   4 marks

b   With reference to atom economy values, explain why the fermentation process
   is becoming popular as a source of ethanol.   2 marks

Question 17 (5 marks)

One of the main methods of producing biodiesel is the base-catalysed transesterification of triglycerides.

Below is a typical example of this process.

a   Calculate the atom economy for this process.   2 marks

b   Describe one potential hazard associated with this process and suggest a way that it
   could be resolved.   3 marks

Question 18 (8 marks)

Ibuprofen is a type of anti-inflammatory drug used in the management of pain relief. It is one of the most
prescribed drugs in the world, with production in the millions of tonnes per year.

The molecular formula of ibuprofen is $C_{13}H_{18}O_2$.

The table below gives information on two different processes used to produce ibuprofen.

	Total	Used in ibuprofen	Unused in ibuprofen
Process A	514.5	206	308.5
Process B	266	206	60

a   With reference to atom economy values, decide which process should be termed 'green'.   4 marks

b   One method of producing ibuprofen is by the Friedel–Crafts acylation of isobutylbenzene followed by
   reduction, chloride substitution and Grignard reaction.

$$C_{10}H_{14} \longrightarrow C_{13}H_{18}O_2$$
isobutylbenzene   ibuprofen

   Calculate the mass of isobutylbenzene required to produce 3800 kg of ibuprofen
   if the yield of this process is 26.5%.   4 marks

Question 19 (6 marks)

The polymer polybutylene terephthalate is an important material with a wide range of applications from electrical wiring insulation to underground pipes.

Its general formula is shown below.

$$\left[\begin{array}{c} O \\ \| \\ C \end{array} \!\!-\!\! \bigcirc \!\!-\!\! \begin{array}{c} O \\ \| \\ C \end{array} \!\!-\!\! O - CH_2 - CH_2 - CH_2 - CH_2 - O \right]_n$$

a   Draw the structures of the monomers used to make this polymer.   3 marks

b   Suggest two differences in properties between this polymer and polypropene.   3 marks

Question 20 (6 marks)

One of the most common forms of fuel cell is the alkaline hydrogen fuel cell.

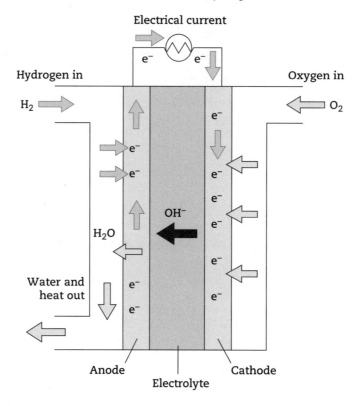

The general equation is:

$$2H_2(g) + O_2(g) \rightarrow 2H_2O(l)$$

Another type is the direct methanol fuel cell, which is similar to the hydrogen fuel cell except that it is methanol, CH_3OH, that is oxidised.

$$CH_3OH(aq) + H_2O(l) \rightarrow 6H^+(aq) + CO_2(g) + 6e^- \quad E° = -0.53 \text{ V}$$

This uses an acid electrolyte so the reduction half-equation is:

$$O_2(g) + 4H^+(aq) + 4e^- \rightarrow 2H_2O(l)$$

a   Compare the standard $E°_{cell}$ values for the hydrogen fuel call and the methanol fuel cell.   4 marks

b   Suggest one advantage and one disadvantage of using a methanol fuel cell over
the hydrogen fuel cell.   2 marks

SOLUTIONS

UNIT 3: EQUILIBRIUM, ACIDS AND REDOX REACTIONS

CHAPTER 1 TOPIC 1: CHEMICAL EQUILIBRIUM SYSTEMS

Multiple-choice questions

Question 1 B

The diagram is for an endothermic reaction, which is the reverse of that shown in the equation. The heat of reaction is, therefore, positive and the arrow goes from reactants to products – arrow 2.

Question 2 D

To force the system to the right (more PCl_5), increasing the pressure shifts the system from the left (2 moles of gas = high pressure) side to the right (1 mol of gas = low pressure) side.

Question 3 D

$$Q = \frac{0.65^2}{0.5 \times 0.25} = 3.38$$

Question 4 A

Strong diprotic acid dissociates completely to produce three ions.

Question 5 D

$pH = -\log_{10}[(0.36 \times 2)] = 0.14$

Question 6 D

The system reduces the number of OH^- ions by increasing the number of ethylammonium ions, which neutralise the OH^-.

Question 7 C

$[H_3O^+] = 4.18 \times 10^{-3}$, $pH = 2.38$

Question 8 B

Strong acid–weak base titration. Equivalence point will be about $pH = 4 = pK_a$ of indicator = bromophenol blue.

Question 9 C

Molar ratio of KHP to NaOH $= 1:1$. $n_{NaOH} = 0.139 \times \left(\dfrac{23.45}{1000} \right) = 3.26 \times 10^{-3} = n_{KHP}$,

$m_{KHP} = 3.26 \times 10^{-3} \times 12.5 \times 204.11 = 8.169$ g

Question 10 C

$14 - 9.37 = 4.63 = pK_a$. $[H_3O^+] = 4.4 \times 10^{-3}$, $pH = 2.4$

Question 11 B

$[CH_3CH_2COO^-] = [H_3O^+] = 1.58 \times 10^{-3}$, % ionisation $= \left(\dfrac{1.58 \times 10^{-3}}{0.185} \right) \times 100 = 0.86\%$

Question 12 C

From first to last, each one loses an H^+ and gains a –

Question 13 A

The forward reaction is endothermic, graph shows system has shifted to the right; therefore, heat must have been added.

Question 14 D

pK_a at the half-equivalence point = 10.3. $pK_b = 14 - 10.3 = 3.7 = -\log_{10}K_b = 2.0 \times 10^{-4}$

Short response questions

Question 15

a

	N$_2$	O$_2$	2NO
Initial	0.15 M	0.15 M	0 M
Change	$-x$	$+x$	$+2x$
Equilibrium	$0.15 - x$	$0.15 - x$	$2x$

$$3.4 \times 10^{-4} = \frac{(2x)^2}{0.15^2}$$

$$[NO] = 2x = 2.77 \times 10^{-3} \text{ M}$$

Mark breakdown
- 1 mark for correctly substituting values into an ICE chart
- 1 mark for determining [NO]

b $K_c(540 \text{ K}) < K_c(440 \text{ K})$

As the temperature increases, K_c gets smaller $\therefore$ [reactants] increase/[products] decrease $\therefore$ if heat were added to the products (exothermic), the reaction would shift left to use up extra heat $\therefore$ reaction must be exothermic.

Mark breakdown:
- 1 mark for correctly relating direction to temperature
- 1 mark for correctly identifying it is an exothermic reaction

Question 16

a In acidic solutions, methyl orange is in its acidic form, structure B, which is red. As base is added, the H_3O^+ ions surrounding the indicator are used up.

At the end point, the H_3O^+ produced by structure B is neutralised. According to Le Chatelier's principle, the indicator shifts to the right to replenish the lost H_3O^+. This produces strucure A, which is yellow.

Mark breakdown:
- 1 mark for correctly identifying structure A as the yellow basic form
- 1 mark for correctly identifying structure B as the red acidic form
- 1 mark for correctly referring to Le Chatelier's principle in explanation

b $K_c \text{ (indicator)} = \dfrac{[In^-][H_3O^+]}{[HIn]}$ At equilibrium [HIn] = [In$^-$]

Therefore, K_c(indicator) = $[H_3O^+]$

Therefore, pK(indicator) = pH

Mark breakdown
- 1 mark for correctly equating [HIn] with [In$^-$] at end point
- 1 mark for correctly equating K_c(indicator) with [H$_3$O$^+$]

Question 17

Moles acid $= 0.08435 \times 1.85 = 0.156 \, \text{mol}$

Moles base $= 0.11345 \times 0.92 = 0.104 \, \text{mol}$

Moles pentanoic acid unreacted $= 0.156 - 0.104 = 0.052 \, \text{mol}$

Concentration of unreacted pentanoic acid $= \dfrac{0.052}{0.08435 + 0.11345} = 0.263 \, \text{M}$

$[H_3O^+] = \sqrt{1.44 \times 10^{-5} \times 0.263} = 1.95 \times 10^{-3} \, \text{M}$

$pH = -\log_{10} 1.95 \times 10^{-3} = 2.7$

Mark breakdown

- 2 marks (1 mark each) for correctly determining the number of moles of the acid and base
- 1 mark for correctly determining the number of moles of unreacted acid
- 1 mark for determining $[H_3O^+]$
- 1 mark for determining pH

Question 18

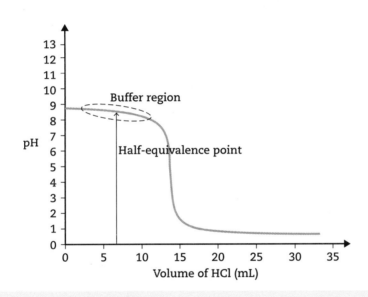

Mark breakdown:

- 1 mark for correctly labelling the buffer region
- 1 mark for correctly labelling the half-equivalence point

b $pK_a + pK_b = 14$

pH at half-equivalence point $= 8.6 = pK_a$

$pK_b = 14 - 8.6 = 5.4$

Mark breakdown:

- 1 mark for determining pK_b

c As strong acid is added, pH changes only slightly:

$CH_3NH_2(aq) + H_2O(l) \rightarrow CH_3NH_3^+(aq) + OH^-(aq)$

The system shifts to the right, producing more OH^- ions which neutralise the extra H_3O^+ added from the HCl.

Mark breakdown

- 1 mark for correctly explaining the operation of a basic buffer solution when acid is added
- 1 mark for using the correct buffer equation

d Equivalence point pH = 4.8.

At the equivalence point, there are no reactants. The products are $CH_3NH_3^+$ and Cl^-. Cl^- = conjugate base of a strong acid, therefore, it has no acid/base properties.

$CH_3NH_3^+$ = conjugate acid of a weak base; therefore, it is a weak acid. pH = 4.8.

Mark breakdown
- 1 mark for correctly identifying the equivalence point (± 0.2)
- 1 mark for correctly identifying the conjugate weak acid and its pH

Question 19

$$K_c(6 \text{ min}) = \frac{0.02^2}{0.116^2 \times 0.066} = 0.45$$

$$K_c(12 \text{ min}) = \frac{0.01^2}{0.13^2 \times 0.085} = 0.069$$

System is exothermic, as defined by the $-\Delta H$.

At $t = 8$ min, system has shifted left; therefore, temperature must have been increased.

This increase in temperature decreased K_c.

Therefore, as temp is increased, K_c decreases

Mark breakdown

1 mark each for correctly
- calculating K_c at 6 min
- calculating K_c at 12 min
- identifying direction shift
- identifying increase in temp
- correct statement – as the temperature increases, K_c decreases

Question 20

Initial moles NaOH = $0.284 \times 0.03 = 8.52 \times 10^{-3}$ mol

Unreacted moles NaOH = $n(HCl) = 0.196 \times 0.01760 = 3.45 \times 10^{-3}$ mol

Moles NaOH that reacted with $NH_4^+ = 8.52 \times 10^{-3} - 3.45 \times 10^{-3} = 5.07 \times 10^{-3}$ mol

$n(NH_4^+) = n(N) = 5.07 \times 10^{-3}$ mol

$m(N) = 5.07 \times 10^{-3} \times 14.01 = 0.071\,g$

$\%N = \dfrac{0.071}{0.504} \times 100 = 14.1\%$

Mark breakdown

1 mark each for correctly determining
- the initial $n(NaOH)$
- the final $n(NaOH)$
- $n(NaOH)$ that reacted with NH_4^+
- $m(N)$
- $n(N)$
- $\%(N)$

SOLUTIONS

UNIT 3: EQUILIBRIUM, ACIDS AND REDOX REACTIONS

CHAPTER 2 TOPIC 2: OXIDATION AND REDUCTION

Multiple-choice questions

Question 1 C

Pb in $Pb(NO3)_2(aq)$ is reduced from Pb^{2+} to Pb, Mn(s) is oxidised to Mn^{2+} in $Mn(NO_3)_2(aq)$

Question 2 D

Electrolytic cell: Anode is positively charged

Galvanic cell: Electrons flow from anode to cathode

Question 3 A

Question 4 B

$E^{\circ}_{cell} = -0.13 + (+0.44) = +0.31$ V

Question 5 C

The pure copper needs to attract the positively charged Cu^{2+} ions; therefore, it is attached to the negative terminal of the power supply.

Question 6 B

$3 \times 2- = 6-$. Remove one − because the ion has a charge of 1 − = 5−. This must be balanced by the iodine = 5+

Question 7 A

$-1.36 + 0.34 = -1.02$ V

Question 8 C

Oxalate is the reducing agent because it loses electrons.

Question 9 D

Cell voltage is negative due to a spontaneous chemical reaction.

Question 10 B

Question 11 C

The strongest reducing agent has the highest negative value/smallest positive value.

Pb = −0.13 V, Ag = +0.80 V, Cl^- = +1.36 V

Question 12 A

Question 13 D

From the electrode potentials table, Fe is more reactive and will be the anode, and Sn will be the cathode.

Question 14 C

As the ocean gets deeper, the amount of dissolved oxygen gets less. The high pressure 'squeezes out' the oxygen.

Short response questions

Question 15

$$16H^+(aq) + 2Cr_2O_7^{2-}(aq) + 3C_2H_5OH(aq) \rightarrow 3CH_3COOH(aq) + 4Cr^{3+}(aq) + 11H_2O(l)$$

OR

Balance half-equations for O and H:

$$14H^+(aq) + Cr_2O_7^{2-}(aq) \rightarrow 2Cr^{3+} + 7H_2O(l)$$

$$H_2O(l) + 3C_2H_5OH(aq) \rightarrow CH_3COOH(aq) + 4H^+(aq)$$

Balance half-equations for charge:

$$6e^- + 14H^+(aq) + Cr_2O_7^{2-}(aq) \rightarrow 2Cr^{3+} + 7H_2O(l)$$

$$H_2O(l) + 3C_2H_5OH(aq) \rightarrow CH_3COOH(aq) + 4H^+(aq) + 4e^-$$

Balance half-equations for charge by multiplying each half-equation by the appropriate factor:

$$12e^- + 28H^+(aq) + 2Cr_2O_7^{2-}(aq) \rightarrow 4Cr^{3+} + 14H_2O(l) \quad (\times 2)$$

$$3H_2O(l) + 3C_2H_5OH(aq) \rightarrow 3CH_3COOH(aq) + 12H^+(aq) + 12e^- \quad (\times 3)$$

Add the half-equations together, ensuring that e^-, H^+ and H_2O are cancelled appropriately:

$$16H^+(aq) + 2Cr_2O_7^{2-}(aq) + 3C_2H_5OH(aq) \rightarrow 3CH_3COOH(aq) + 4Cr^{3+}(aq) + 11H_2O(l)$$

Mark breakdown

- 4 marks for giving the correctly balanced equation

OR

- 1 mark for correctly balancing for O and H
- 1 mark for adding the correct number of e^- to both half-equations
- 1 mark for multiplying each species in the half-equations by the correct factor
- 1 mark for adding the half-equations and correctly cancelling e^-, H^+ and H_2O

Question 16

a L is oxidised $\rightarrow L^{2+}$; therefore, it must be the reductant.

Mark breakdown

- 1 mark for correctly identifying L as the reductant

b Q nitrate formula = $Q(NO_3)_2$

$NO_3^- = -1$

Q must be +2

Mark breakdown

- 1 mark for correctly identifying the oxidation state as +2

c

Oxidised species ⇌ Reduced species	$E°$ (V)
$Q^{2+}(aq) + 2e^- \rightarrow Q(s)$	−0.24
$T^{2+}(aq) + 2e^- \rightarrow T(s)$	−0.13
$L^{2+}(aq) + 2e^- \rightarrow L(s)$	+0.14
$D^{2+}(aq) + 2e^- \rightarrow D(s)$	+0.73

Mark breakdown
- 2 marks for correctly matching the letter with the standard electrode potential
- 2 marks for correct half-equations
- 1 mark for explaining that more reactive metals have larger negative values

Question 17

a Anode half-equation: $2H_2(g) + 4OH^-(aq) \rightarrow 4H_2O(l) + 4e^-$

Cathode half-equation: $O_2(g) + 2H_2O(l) + 4e^- \rightarrow 2H_2O(l)$

Anode product $= H_2O(l)$

Cathode product $= OH^-(aq)$

Mark breakdown
- 2 marks (1 mark each) for correctly identifying the half-equations
- 2 marks (1 mark each) for correctly identifying the products

b $E°_{red} = +0.40$ V

$E°_{ox} = (\text{reverse} -0.83) = +0.83$ V

$E°_{cell} = +0.40 + (+0.83) = +1.23$ V

Mark breakdown
- 1 mark for correctly identifying red $E°$ value
- 1 mark for correctly identifying ox $E°$ value
- 1 mark for correctly calculating $E°_{cell}$

c $2H_2(g) + O_2(g) \rightarrow 2H_2O(l)$

Mark breakdown
- 1 mark for correctly determining the overall equation

Question 18

a 1 = cathode

2 = anode

Mark breakdown
- 2 marks (1 mark each) for correctly identifying cathode and anode

b Oxidation: $2Cl^-(l) \rightarrow Cl_2(g) + 2e^-$

Reduction: $Na^+(l) + e^- \rightarrow Na(l)$

Mark breakdown
- 2 marks (1 mark each) for correctly identifying both half equations

c $E^\circ_{red} = -2.71$ V

 $E^\circ_{ox} = (\text{reverse} +1.36) = -1.36$ V

 $E^\circ_{cell} = -2.71 + (-1.36) = -4.07$ V

Mark breakdown

- 3 marks (1 mark each) for correctly identifying $E^\circ_{red,}$ $E^\circ_{ox,}$ E°_{cell}

Question 19

a Anode: $Cu(s) \rightarrow Cu^{2+}(aq) + 2e^-$

 Cathode: $Fe^{3+}(aq) + e^- \rightarrow Fe^{2+}(aq)$

Mark breakdown

- 2 marks (one mark each) for correctly identifying both half-equations

b $Cu(s) + 2Fe^{3+}(aq) \rightarrow Cu^{2+}(aq) + 2Fe^{2+}(aq)$

Mark breakdown

- 1 mark for correctly determining the overall equation

c $E^\circ_{red} = +0.77$ V

 $E^\circ_{ox} = (\text{reverse} +0.34) = -0.34$ V

 $E^\circ_{cell} = +0.77 + (-0.34) = +0.43$ V

Mark breakdown

- 1 mark for correctly identifying E°_{ox}
- 1 mark for correctly calculating E°_{cell}

Question 20

a Any metal higher than iron in the standard electrode potential table (see page 62)

 A metal higher than iron in the table is more reactive and so will oxidise in preference to iron and so has a larger negative standard electrode potential value than iron.

 e.g. Zn ($E^\circ = -0.76$ V)

 Fe ($E^\circ = -0.44$ V)

Mark breakdown

- 1 mark for correctly identifying a suitable metal
- 1 mark for providing a suitable explanation

b Sample answer:

 e.g. $Zn(s) \rightarrow Zn^{2+}(aq) + 2e^-$

 When Zn oxidises, it produces electrons.

 The electrons travel through the plastic-coated wire to the iron.

 The electrons prevent iron from oxidising/corroding.

Mark breakdown

- 2 marks for providing a suitable explanation

c   Possible answers include:

The pipeline could be painted. This prevents O_2 and H_2O oxidising the iron.

OR

The pipeline could be electroplated with another metal. This prevents O_2 and H_2O oxidising the iron.

OR

An electric current could be supplied to the pipeline.

Mark breakdown
- 2 marks for providing a suitable answer and explanation

SOLUTIONS

UNIT 3: EQUILIBRIUM, ACIDS AND REDOX REACTIONS

CHAPTER 3 DATA TEST

Data set 1

Question 1

The concentration of SO_2 changes from 0.092 M to 0.128 M, while the concentration of SO_3 changes from 0.022 M to 0.004 M. (Allowable range $\pm$ 0.002 M)

Mark breakdown
- 1 mark for identifying each correct concentration change

Question 2

At 2 minutes, some SO_3 was removed. The decrease in the other gases is because the system has shifted to the right, which means that their concentrations will decrease.

At 12 minutes, there was a temperature change – all concentrations changed gradually.

SO_3 decrease, SO_2 and O_2 increase means that the system has shifted to the left. According to the thermochemical equation, this is the endothermic direction. Therefore, the temperature must have been increased.

Mark breakdown
- 1 mark for correctly identifying concentration change at 2 minutes
- 1 mark for correctly identifying temperature change at 12 minutes
- 1 mark for correctly stating increase in temperature

Question 3

K_c at 11 minutes: $K_c = \dfrac{(0.01)^2}{(0.11)^2 \times 0.06} = 0.1377 = 0.138$

K_c at 17 minutes: $K_c = \dfrac{(0.004)^2}{(0.128)^2 \times 0.079} = 0.01236 = 0.0124$

The change at 11 minutes indicates an increase in temperature. Therefore, at 17 minutes the system is at a higher temperature. Its K_c value is smaller.

Therefore, the higher the temperature, the smaller the K_c.

Mark breakdown
- 1 mark for correctly calculating K_c at 11 and 17 minutes (2 marks total)
- 1 mark for correctly stating higher temperature at 17 minutes
- 2 marks for relating a higher temperature to a great K_c value

Data set 2

Question 1

pH at equivalence point is 7.8.

Mark breakdown
- 1 mark for correctly identifying the equivalence point

Question 2

Volume of NaOH at equivalence point is 15.5 mL, volume NaOH at half-equivalence point $= \dfrac{15.5}{2} = 7.75$ mL

OR

The volume of NaOH at the half-equivalence point is 7.75 mL, which is half the volume at the equivalence point. pH at half-equivalence $= 4.3 \pm 0.02$

AND the pH at the half equivalence point is lower than the pH at the equivalence point.

Mark breakdown
- 1 mark for correctly distinguishing the half-equivalence point from the equivalence point
- 1 mark for reference to pH
- 1 mark for reference to volume of NaOH added

Question 3

pH at half-equivalence point $= pK_a$ of the acid, 4.3.

Using $pK_a = -\log_{10} K_a$

Therefore, $K_a = 5.01 \times 10^{-5}$

The closest value from the table is benzoic acid.

Mark breakdown
- 1 mark for correctly identifying pK_a from the curve
- 1 mark for correctly calculating K_a
- 1 mark for correctly identifying benzoic acid

Question 4

Moles of NaOH added from burette = moles of unknown acid at equivalence point

$= 0.103 \times \dfrac{15.5}{1000} = 1.5965 \times 10^{-3}$ mol

Concentration in the 20.00 mL sample $= \dfrac{1.5965 \times 10^{-3}}{\dfrac{20.00}{1000}} = 0.0798 = 0.08$ M

Taking the dilution into account $= 0.0798 \times 20 = 1.596 = 1.60$ M

Mark breakdown
- 1 mark for correctly calculating moles at equivalence point
- 1 mark for correctly calculating concentration
- 1 mark for accounting for dilution

SOLUTIONS

UNIT 4: STRUCTURE, SYNTHESIS AND DESIGN

CHAPTER 4 TOPIC 1: PROPERTIES AND STRUCTURE OF ORGANIC MATERIALS

Multiple-choice questions

Question 1 D

The compound contains the ester functional group.

Question 2 C

A large, broad peak at 3500 = O–H, peak at 1000 = C–O = alcohol

Question 3 B

The compound has molecular formula C_6H_{14}. Option B is the only one with the same formula. All the others have molecular formula of C_7H_{16}.

Question 4 D

These are in the correct order of decreasing intermolecular forces strength.

Question 5 D

The I–I adds across the double bond between C_2 and C_3.

Question 6 C

Question 7 C

Reaction with permanganate indicates a primary or secondary alcohol. No reaction with sodium carbonate indicates that the product is a ketone, so X must be a secondary alcohol.

Question 8 C

The central N atom has an ethyl group and a methyl group attached to it. The group containing the C=O has four carbon atoms: butanamide.

Question 9 B

Asn has isoelectric point = 5.4. In a pH 7.8 buffer, it loses its H and becomes negative. It is repelled by the cation exchange resin, which is also negative.

Question 10 A

Question 11 A

The base peak is the one that has the highest abundance = 43.0

Question 12 A

Question 13 A

Question 14 D

The longest carbon chain is 4–nonene, which goes across the double band, *trans*. Ethyl group at C_4, methyl at C_5

Short response questions

Question 15

Compound X must be an *alkane*. Whenever UV light is used with a halogen the reaction is a substitution. The product is Q, a *chloroalkane*.

T must be an *alcohol* because when a chloroalkane is reacted with aqueous NaOH, a nucleophilic substitution reaction occurs in which an –OH group replaces the chlorine atom.

Compound L must be a *carboxylic acid* because it is the product of the oxidation of an alcohol AND when mixed with Na_2CO_3 solution, bubbles of colourless gas were produced.

> **Hint**
> The production of bubbles of colourless gas when mixed with an unknown substance is a test for carboxylic acids. This is a 'carbonate + acid' reaction and the colourless gas is CO_2.

If L is a carboxylic acid, then T must be a *primary* alcohol.

Compound N must be a *primary* amine because it is the product a nucleophilic substitution reaction between a chloroalkane and ammonia.

The chlorine atom is at the end of the carbon chain in T because it produced a primary alcohol.

Compound G must be an *amide* because it is the product of the reaction between L (carboxylic acid) and N (amine).

If G has a molecular mass of 143, it must be *N-butyl butanamide*.

G

$$CH_3CH_2CH_2 - \overset{\overset{\displaystyle O}{\|}}{C} - \underset{\underset{\displaystyle H}{|}}{N} - CH_2CH_2CH_2CH_3$$

N-butyl butanamide

N

$CH_3CH_2CH_2CH_2NH_2$

1-butanamine

Q

$CH_3CH_2CH_2CH_2Cl$

1-chlorobutane

L

$CH_3CH_2CH_2COOH$

Butanoic acid

T

$CH_3CH_2CH_2CH_2OH$

1-butanol

X

$CH_3CH_2CH_2CH_3$

Butane

Mark breakdown
- 1 mark each for correct name *and* structure (6 marks total)
- 4 marks for explanation of structures and reactions

Question 16

a

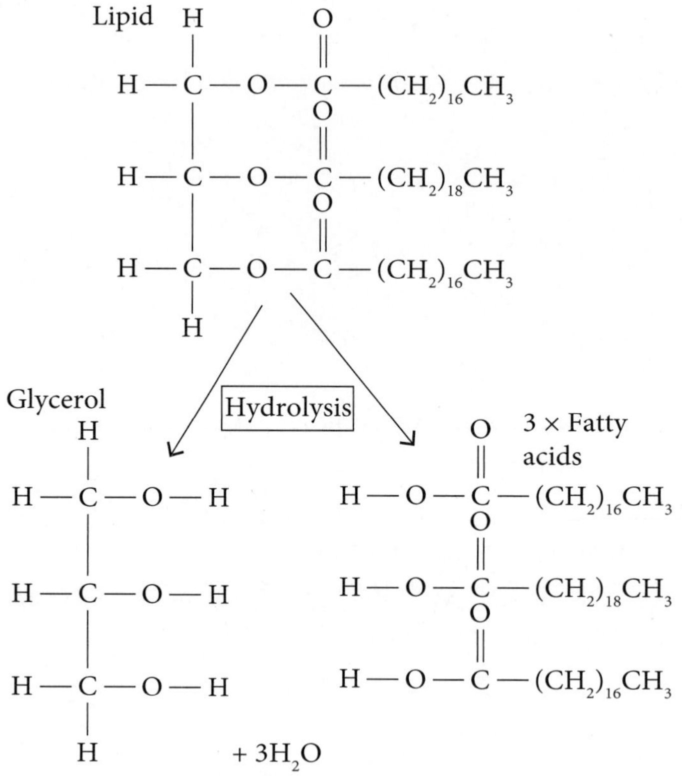

Mark breakdown

- 1 mark for correctly showing the molecule splits
- 1 mark for correctly labelling the hydrolysis
- 1 mark for correctly labelling the reactants glycerol and three fatty acids

b

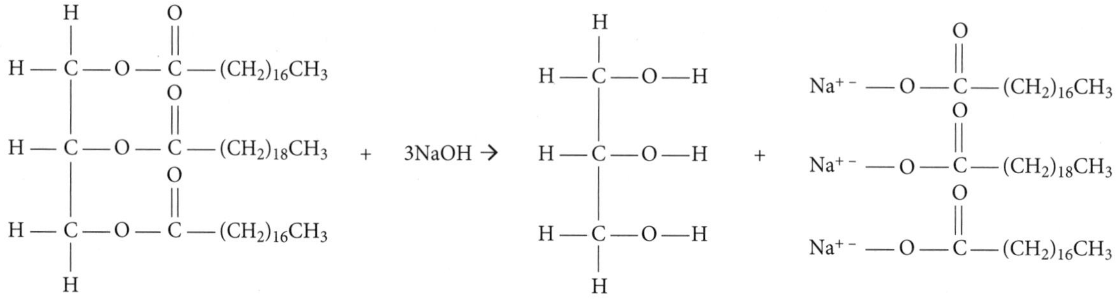

Hydrolysis   Glycerol   Sodium salts of fatty acids

Mark breakdown

- 1 mark for each correct structure

Question 17

a   **A:** N-Ethyl-N-methylbutanamide

   B: N-Ethylethanamide

   C: Ethanamide

Mark breakdown

- 1 mark for each correctly named substance

b **A** = Tertiary amide: R = CH$_3$CH$_2$CH$_2$, R' = CH$_3$, R" = CH$_3$CH$_2$

   B = Secondary amide: R = CH$_3$, R' = H, R" = CH$_2$CH$_3$

   C = Primary amide: R = CH$_3$, R' = H, R" = H

Mark breakdown
- 1 mark for each correct identification of amide (3 marks total)
- 1 mark for correct identification of each R group (3 marks total)

Question 18

a Predominant intermolecular force = H-bonding

1-butanol has higher melting point; therefore, it has stronger intermolecular forces. The –OH group is at the end of the chain. Therefore, the dispersion forces between the uninterrupted carbon chain augment the H-bonding.

2-Butanol is much more soluble than 1-butanol.

Both have H-bonding, as does water. The long, uninterrupted hydrocarbon chain in 1-butanol will have difficulty separating water molecules.

2-Butanol has the –OH group in the middle of the chain. The chain is interrupted; therefore, the molecule is more soluble.

Mark breakdown
- 1 mark each for correct discussion of melting points and solubility (2 marks total)

b Predominant intermolecular force is dipole–dipole interaction. Both molecules have the C=O group. Butanal is at the end of the carbon chain.

Butanone is in the middle of the chain.

There is very little difference in boiling points but the C=O group in butanone is slightly stronger and so the dipole attractions between ketone molecules is slightly greater.

Butanone is much more soluble in water than butanal. This is due to the long, uninterrupted hydrocarbon chain in butanol having trouble separating water molecules.

Mark breakdown
- 1 mark each for correct discussion of melting points and solubility (2 marks total)

Question 19

The two types of monosaccharide are aldose sugars that have an aldehyde group and ketose sugars that have the ketone group:

   Aldose sugars: A, B, E

   Ketose sugars: C, D, F

Mark breakdown
- 2 marks for recognising the two types of monosaccharide
- 4 marks for correctly identifying each sugars' group

Question 20

a   Threonine pI = 5.6, aspartic acid pI = 3.0, buffer = 4

threonine + buffer → threonine⁻

aspartic acid + buffer → aspartic acid⁺

Once injected into the well, the threonine will move towards the positive side and aspartic acid will move towards the negative side.

This is how they are separated.

Mark breakdown
- 1 mark for each correct formula (2 marks total)
- 1 mark for each correct conclusion

b

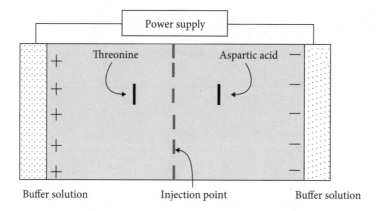

Eventually, each amino acid will come to a part of the gel that has the same pH as their respective pI. They will become neutral and therefore insoluble in the gel and will precipitate out.

Mark breakdown
- 1 mark each for threonine and aspartic acid being on the correct side. Actual position is not important.
- 1 mark for correct explanation

SOLUTIONS

UNIT 4: STRUCTURE, SYNTHESIS AND DESIGN
CHAPTER 5 TOPIC 2: CHEMICAL SYNTHESIS AND DESIGN

Multiple-choice questions

Question 1 B

Question 2 C

$$AE = \left(\frac{183.9}{58 + (2 \times 126.9)} \right) \times 100 = 58.99\%$$

Question 3 C

Question 4 A

Isomers that are chiral cannot be superimposed on their mirror images by any kind of rotational or translational transformation.

Question 5 A

Option B produces hydrogen, option D doesn't do anything, option C is quite rare.

Question 6 A

A high AE means that most of the reactants are converted to desirable product with very little waste.

Question 7 C

Question 8 D

Oxygen is reduced at the cathode and hydrogen is oxidised at the anode.

Question 9 D

Glycogen is similar to starch – it is a polymer of α-glucose.

Question 10 B

Question 11 B

Question 12 A

High pressure forces the reaction to the right-hand side (low pressure side). Although high temperature forces the reaction to the left, the resulting higher rate of reaction compensates for this.

Question 13 A

Question 14 A

Short response questions

Question 15

a

$$CH_3 \qquad CH_3 — CH — CH_3$$
$$| \qquad\qquad\qquad |$$
$$H_2N — CH — CO — N — CH — COOH$$

Amino group   Peptide link | Carboxyl group
$$H$$

Name: Valala

Mark breakdown

- 1 mark for correctly identifying the amino acids required
- 1 mark for correct attachment

b See part a

Mark breakdown

- 1 mark each for identifying type of bond, position and name of both functional groups and dipeptide

Question 16

a Fermentation: atom economy $= \dfrac{2 \times 46.08}{180.18} \times 100 = 51.15\%$

Ethene hydration: atom economy $= \dfrac{46.08}{46.08} \times 100 = 100\%$

Mark breakdown

- 2 marks for each correct calculation

b Fermentation is becoming popular even though its atom economy is only 51.15% compared to the 100% for hydration of ethene because ethene is NOT green – it comes from crude oil (or coal or gas).

Fermentation uses a sustainable crop.

Mark breakdown

- 1 mark for correctly stating atom economy values
- 1 mark for stating ethene from oil comes from fossil fuels (oil, coal or gas)

Question 17

a Atom economy $= \dfrac{931.74}{927.27 + (3 \times 32.05)} \times 100 = \dfrac{931.74}{1023.85} \times 100 = 91.0\%$

Mark breakdown

- 2 marks for correctly calculating the answer

b Possible answers include:

- Excess heat is generated, which leads to problems with cost and safety. Could lower the rate of production, which would reduce the heat produced.
- Long-chain fatty acids in contact with sodium hydroxide could produce soap, which will lower the efficiency of the process. Could use a biological enzyme such as lipase instead of NaOH.

Mark breakdown
- 1 mark for correctly identifying a hazard
- 1 mark for accurate explanation of the problem
- 1 mark for a potential remedy

Question 18

a Process A atom economy $= \dfrac{206}{514.5} \times 100 = 40.0\%$

Process B atom economy $= \dfrac{206}{266} \times 100 = 77.4\%$

Process B is green. It has a higher atom economy.

Therefore, there is less cost and less waste.

Mark breakdown
- 1 mark for each correct percentage
- 1 mark for correctly identifying which process is greener
- 1 mark for linking greenness with atom economy value

b % yield $= 26.25\% = \dfrac{\text{Actual mass}}{\text{Theoretical mass}} \times 100$

Theoretical mass $= \dfrac{3800}{0.265} = 14\ 340$ kg

∴ Number of moles of isobutyl benzene required $= 69\ 514$ mol

∴ Mass of isobutyl benzene required $= 69\ 514 \times 134.22 = 9330$ kg

Mark breakdown
- 1 mark for identifying the correct equation
- 2 marks for correct rearrangement of the equation
- 1 mark for the correct final answer

Question 19

a

Mark breakdown
- 1 mark for correctly identifying the location of the bond
- 1 mark for each correct structure

b Polypropene will have larger side groups, therefore it will:
- be less dense
- have a lower melting point
- be softer.

Mark breakdown
- 1 mark for identifying polypropene will have larger side groups
- 1 mark each for two correct properties

Question 20

a. E_{cell} for alkaline H_2 cell = $E_{red} + E_{ox}$ = +0.4 + (reverse −0.83) = + 1.23 V

E_{cell} for methanol cell = $E_{red} + E_{ox}$ = +1.23 + (−0.53) = + 0.70 V

The alkaline H_2 cell produces 0.5 V more power than the methanol cell.

Mark breakdown
- 1 mark for each correct calculation (3 marks total)
- 1 mark for correctly stating which cell produces the higher voltage

b Advantage: Methanol is easy to produce and transport in large quantities.

Disadvantage: It delivers less power.

Mark breakdown
- 1 mark each for an advantage and a disadvantage